May 1993

Seed-Production Mechanisms

Proceedings of a workshop held in Singapore, 5–9 November 1990

Edited by
Neil Thomas and Nicolas Mateo

INTERNATIONAL DEVELOPMENT RESEARCH CENTRE
Ottawa • Cairo • Dakar • Johannesburg • Montevideo • Nairobi • New Delhi • Singapore

ISBN 0-88936-693-4

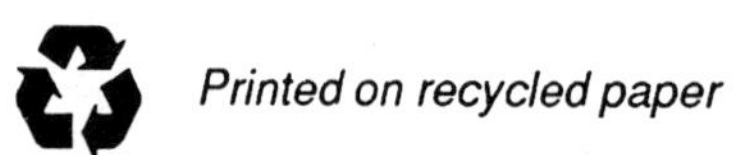

TABLE OF CONTENTS

SEED PRODUCTION MECHANISMS MEETING

Introductory Comments

Chris McCormack, Acting Regional Director
ASRO, IDRC, Singapore

1. Welcome to IDRC's Regional Office. I am sure that Nicolas has made proper and adequate arrangements, however, if you feel we can assist in additional ways during your stay, please let us know.

2. Research on crop improvement has always been, and remains, a cornerstone of IDRC's support to agricultural research. As a former AEP program officer myself, I believe that increased productivity in crop production remains as a prime determinant for poverty alleviation. As both fertile land and inorganic resources become increasingly scarce - and therefore increasingly costly - continued and increased exploitation of the genetic potential of the available germplasm is essential. Growth in agricultural output and agricultural incomes (as for growth from all sources) will have to be less resource-intensive over time. Failure to do so will result in at least two major problems for the people of developing countries: **first**, an aggregate decrease in per capita food availability, and, **second**, an economic `marginalization' of resource-poor farm households - i.e. an inability to compete for national/regional expenditures on food. Both of these `results' are very serious threats to the social and political stability of a country.

   However, resource productivity in itself is insufficient to prevent such economic marginalization over time. The `outputs' of crop genetic improvements must also be related to the market preferences for specific crop characteristics. In other words, crop improvement must also result in value-added per unit of crop produced, as well as providing this value-added at lower cost. In economic terms, crop improvement must be simultaneously `supply-push' (increased output at lower cost) **and** `demand-pull' (command market prices for preferred crop qualities). Such a two-pronged strategy allows for stabilization/increasing farm household incomes **and** consumer satisfaction. There do exist interdisciplinary (economic, cereal chemistry and statistics) research methodologies that allow the definition and prioritization of preferred crop characteristics, and therefore a `guide' for subsequent genetic research. Recent research at IRRI, and in selected national programs, supported by IDRC, provides an example of this interdisciplinary research and the use of its outputs.

3. IDRC recognizes the importance of case study reviews of past research as an important methodology for identifying the determinants and constraints to the successful use of crop improvement activities. We are confident that this meeting, especially given the participants involved, will make a further important contribution to increasing the efficacy of crop research.

4. I wish you all much success, and look forward to the results of your meeting. Thank you.

SEED PRODUCTION MECHANISMS MEETING: SOME ISSUES FOR IDRC

Introductory Comments

Nicolas Mateo, Associate Director (Crop Production Systems)

The development of improved crop cultivars has been an effective way of increasing agricultural productivity of selected crops. Impact achieved is not only dependent on improved germplasm but often on inputs and marketing outlets. Wheat, rice and several plantation crops are perhaps the best known examples. New cultivars do not necessarily require additional inputs by farmers, for example in the case of disease resistance substantial benefits may accrue simply by substituting a new for an old cultivar.

Recognizing the potential contribution of crop genetic improvement, IDRC has given a high priority to supporting national and international plant breeding programs in all regions. These are long-term activities, normally requiring 10 or more years from the identification of a problem or opportunity, to the use of a new cultivar by farmers. A number of reports show that several IDRC-supported projects are now having a positive impact at the farm level. In spite of such reports there are also strong indications that in many projects impact has been limited due to the lack of adequate mechanisms and policies for the dissemination to farmers of seed or other planting materials.

The Agriculture, Food & Nutrition Sciences Division of IDRC has given highest priority to supporting crop improvement research on species which tend to be grown by small-scale, resource-poor farmers, these species are generally neglected by other donor and research organizations. Thus sorghum and millets have received far greater support from IDRC for breeding than have rice, wheat or maize. Although this strategy may be considered rational in terms of resources allocation, the dissemination of the results of such research is often more difficult than in the case of major crops for which effective government or private schemes are already in place.

This meeting considers eight IDRC-supported breeding projects which have, to a greater or lesser extent, successfully disseminated improved planting materials to farmers. The studies cover a range of species, regions and dissemination mechanisms. It is expected that constraints and limitations to wider dissemination and adoption will be important components of the discussions and conclusions.

There are other IDRC-supported projects and activities which will not be represented at this meeting. Many of them could be considered as "building blocks" of larger edifices in which national programs are making long-term crop improvement investments, and therefore it is too early or too complex to get a clear picture of impact or the lack of it. Some of these projects yield advanced lines that are then used for further breeding and improvement by research teams before distribution to farmers is accomplished.

One project not represented here is Improved Crops (Chile). Selected local potato cultivars have been genetically improved by crosses and then given back to the same farmers where the collections took place for further use and evaluation. Conclusive results are not yet published, however, it appears that the Project has become a useful dissemination mechanism and constitutes a novel approach.

We are fortunate to have a participant from CIAT. This Center's valuable experiences in artisanal seed production will be relevant to all other researchers.

It seems to be essential and a priority for national programs (and certainly for IDRC) to learn and understand more about policies and seed production and distribution mechanisms. This knowledge should guide us and modify, if necessary, future actions and interventions.

There are other aspects, closely related to the objectives of this meeting, which will not be part of the discussions but that will certainly affect research and development in the future. The issue is intellectual property rights and the idea of highlighting this now is to put into proper perspective the continuum of seed origins, improvement and evaluation, production, distribution, and utilization systems.

In many developed countries, crop varieties are protected by Plant Breeders Rights, which provide breeders exclusive rights for marketing the varieties, but allow free use for further breeding. Farmers may also multiply seeds for replanting in the next season (this is known as farmer's exemption). The new emerging patent laws for genetic materials may prevent the use of patented genetic information in further breeding, and may also disallow farmer's exemptions.

The other key issue is biodiversity. It is often claimed that plant breeding and PBR decrease biodiversity by encouraging the production and dissemination of new varieties which often replace the more diverse landraces and local crops. Take for example the rich variation of potatoes found in the Andean region. This came about as a result of the many micro-environments found there and the patient and long-term selection efforts of the native communities. Unfortunately there are not clear-cut incentives or compensation for the millions of small growers who have manage to develop, maintain, and make available to humankind this impressive legacy. A World Fund for the global conservation of these resources (including ex-situ formal gene banks as well as in-situ conservation and utilization of genetic resources at the community level) has now been proposed by several individuals and organization.

Seed production and distribution mechanisms are in brief our main concern during this meeting, however we must be prepared to face in the near future a new round of issues that will have considerable influence in research and development in the poorer countries.

SEED PRODUCTION AND DISTRIBUTION MECHANISMS

A Review Paper

Neil Thomas, IDRC Study Coordinator*

Abstract

Crop improvement through breeding is a principal strategy in improving food security for resource-poor farmers. However, farmers do not necessarily adopt the new variety or technology made available to them. Is this because of a lack of awareness of potential benefit, or because the technology does not fit the farming system? Do farmers use different evaluation criteria from those used by breeders?

Formal dissemination assumes that adoption of a new technology is desirable, and that there will be a measurable benefit. While it is possible to define dissemination mechanisms and their components, different degrees of formality among them almost certainly mean that no single one would fully account for the adoption noted. Few programs or projects attempt to quantify the contribution of different mechanisms to dissemination, or, ultimately, adoption by the farmer.

Several projects reviewed note the unique circumstances of each that contributed to, or constrained, successful dissemination. Involvement of the farmers themselves in seed production is often a key to success, including the establishment of community-level seed businesses.

Successful donor-funded seed programs have been characterized by good management, prior experience with seed, and good demand for seed. Rigid government controls are likely to limit success in production and dissemination. Seed programs for marginal agricultural populations require a greater level of effort than those targeted at commercial operations.

1. Introduction

Seed is fundamental to agriculture. It is both the means of transference of genetic information from one crop generation to another, and the basis of economic yield of the majority of crops. Through selection processes practised by farmers over centuries, many crops have become adapted to specific growing conditions, and have evolved qualitative and other characteristics that mirror the preferences of the growers. It is only in the last century that plant breeding has become a scientific process largely out of the hands of the farmers. This has resulted in large-scale changes in the characteristics of some crops, and in the methods of seed production for future crops. Agriculture in many countries is no longer a cyclical process that is contained within the boundaries of the farm.

Plant breeding has developed at the same time as other crop sciences. Much of the improvement attained through breeding and selection is dependent on other agronomic elements, such as fertilization and pest control. Thus, while the seed contains the potential for improved crop yield (or whatever the breeding objective was), the potential may not be achieved without other inputs. Most plant breeders test their crosses and selections under specific conditions. These are often the conditions recommended to farmers for the management of their crop. However, the closely controlled conditions of test plots are generally not duplicatable on-farm, and crop output will vary according to within and between farm conditions.

* Thomas Development Associates Ltd., Lostwithiel, Box 58, R.R.#1, Mallorytown, Ontario, Canada K0E 1R0

Very often, therefore, the crop does not achieve the potential described for it by the breeder.

The resource-poor small farmer is even more at the mercy of natural forces. Normally unable to provide extra inputs, the farmer's system would revolve around whatever natural fertility is present in the soil, plus whatever the farmer can return to it in the form of animal and crop residues. Variable weather and pest cycles will interact with these resources for the overall definition of growing conditions. Under such uncertain conditions, the farmer will need very strong convincing of the desirability of changing his or her crop or crop variety. This small farmer is thus normally conservative, with a range of reasons for retaining current varieties or production methods. In such conservatism is security.

2. Our Hypothesis

As breeders or other crop scientists, we believe that improving the characteristics of a crop is one of the main ways of improving food security and general welfare of individual farm families and communities, whether the crop is grown for home consumption, or some is to be sold. Through the Green Revolution we have seen the potential of this approach, where the output of crops such as rice and wheat have increased significantly. Yet this again required the provision of other inputs, something of which not all farmers were capable, or to which they may not have aspired. As a result, the poorer farmers have tended to become marginalized, even though there is a technological basis for increasing crop production.

In many cases, the market may not be equivalent to the potential of the technology, i.e., prices paid for the crop do not cover the input costs, especially where marked increases in yields depress prices. In such situations, lower-cost production alternatives are usually sought, with concomitant reduction in yield. However, institutions may still recommend production practices that assume little or no marketing constraint. Small farmers who accept production credit under such conditions run the risk of not being able to repay, and incurring long-term indebtedness.

3. Adoption

If adoption can be described as the uptake of a new technology by a producer, the fact that farmers, especially small farmers, often do not immediately adopt technology on a wide-scale (e.g. Anon, 1987) poses a problem for crop scientists. Often this is assumed to be a consequence of the intractability of the peasant, who, because of poor education, knows no better. However, there are many factors at play in this process, and it is often the case that small-farm communities do have persons willing to adopt new technology (Fairlie et al., 1990), but who are cognizant of factors not necessarily clear to the scientist behind the technology, factors that may nullify any advantage in adoption.

It is not always easy to measure adoption, because this is not a one-time affair. Thus, a new variety may be tried one year, but, for different reasons, may not appear on the same farms each year over successive years. The farmer's production resources may change from year to year, forcing or requiring some change in what is grown and how it is grown. However, the farmer will be cognizant, if the extension services have done their work, of what is available to him, and how to use it.

One measure of adoption is the index of acceptability, which is calculated as the percentage of farmers who continue to use the technology (A), multiplied by the percentage of their crop on which the technology is applied (C), over 100:

$$IA = \%A \times \%C \ / \ 100$$

This index has been helpful in evaluating the acceptability of new technology (Dardon, 1982), though it should only be used on farms which conducted tests, i.e. it should not be extrapolated to a wider population. This author notes that in Guatemala an index of 25 in regions of traditional agriculture was considered good for the adoption of improved maize varieties. An index of this type offers a means of quantifying adoption, though explains none of the underlying reasons (which may be both for and against); it should be noted that an identical value for IA could be calculated by reversing the values for A and C.

Adoption, then, is one of the factors that anyone involved in technology development must consider closely. Current farming-systems methodology includes technology testing and validation as key components of the transfer process. Plant breeders are no less responsible for assuring themselves of the appropriateness of their outputs than is any other agricultural scientist. If a farmer appears indifferent to a new variety, what is the reason? Is it a question of not understanding potential advantages of the new variety? Is there an aspect of changing varieties of which the breeder is unaware, but which to the farmer represents an obstacle? Is the farmer using different evaluation criteria than the breeder, such that the material is seen differently? The literature is full of references to lack of adoption, yet the reasons are rarely elucidated.

While farming-systems approaches have been widely adopted in, for instance, the development of improved agronomic practices, there has been more resistance to their inclusion in breeding programs and varietal development. Yet there is evidence (e.g. Wooley, 1986) that such an approach is no less critical in the latter activities. Some work in Colombia indicated that the order of superiority among ten improved lines of beans was very different between on-station and on-farm tests, and that local conditions could be such that there was no correlation between yields in the two cases. In his writing, Wooley effectively asks the question `what would the on-farm yields have been of the varieties eliminated in the process of selecting the ten elite on-station lines?' Even if this latter concern is ignored, to what extent would a solely on-station program have resulted in potentially lower levels of adoption of a new bean variety? Would a breeding program conducted primarily on-farm result in a significantly different result from one almost wholly station-oriented?

Farmer evaluation and adoption can speed up the process of varietal release. Wooley (1986) again cites the case in Colombia where farmers were already disseminating seed of a bean line under test by the third year of such testing. This acceptance, accompanied by evidence of significant increases in yield and disease resistance when compared to the traditional variety, forced the breeding institution into immediate formal release. Cock (1986) suggests that farmers are capable of managing a low-cost trial network, an innovation that would certainly add much debate to the evaluation process.

Douglas (1980) has attempted to list in generic terms the factors that farmers consider when deciding whether to adopt a new variety. These are over and above the institutional and market factors that may influence seed availability and quality. He describes them as follows:

1. **Relative advantage.** This is the degree to which the new variety will raise benefits, lower costs, compared to benefits or costs associated with current varieties. It may also take the form of a difference in effort, risk, prestige or social approval.
2. **Reliability.** The new variety will consistently produce the minimum crop needed to feed his family and provide the income normally received from sales.
3. **Simplicity.** The ease of use.
4. **Compatability.** With respect to needs, values, past experience and the farming system.
5. **Visibility.** With respect to the results of the new variety in the eyes of the farmer and others.
6. **Divisibility.** The perception that the innovation can be tried on a limited basis.
7. **Independence.** That the variety can be adopted without consulting anyone else.

It is not clear whether such an analysis has any practical value in promoting adoption, though the first two points obviously reflect on the economic advantage conferred (through both increased returns and stability of return). The other points reflect more on socio-cultural issues which may influence adoption. Perceptions of the intended recipient can be very significant - Douglas (op. cit.) again cites an instance where the perception of attributes explained more than half of the variance in the rate of adoption.

The issue of adoption suggests that the breeder must be more than just a breeder. Experience is needed in on-farm evaluation techniques, in knowing how to interact with farmers, in understanding local marketing issues (whether of seed or the harvested crop), in crop processing, in the cultural and social characteristics of the target population, etc.. If the breeder has not the social science skills necessary for some of these areas, then crop improvement becomes an inter-disciplinary effort, involving more than one scientist. In other words, the research effort requires more than the release of a new variety.

Then, of course, once the farmer is convinced that he or she wants to use the new variety, it must be accessible. The seed must be available.

4. Dissemination Mechanisms

The dissemination process could be considered to have two components, the apprising of the farm community of the new technology (hopefully, the involvement of the community in its development), and the delivery of the new technology. It is possible for a single channel to serve both functions. Where large volumes of material are involved, there may be steps in the dissemination process which do not involve the farming community, but which eventually target this group. Equally, there may be mechanisms which impose controls on this process, either to protect the end user, or even the originator of the technology.

Dissemination assumes that adoption of a new technology is desirable, that there will be a measureable benefit. Much of the emphasis will therefore be on the characteristics of the technology that will impart this benefit. Traditional approaches centred on demonstration techniques, which were considered to show the advantages to be gained by the new technology. To the scientist, this was often a black-and-white issue, with perhaps a single indicator, e.g. crop yield, being used as the basis for decision-making. However, traditional farmers may have had a multitude of factors to take into account, rendering an evaluation much more complex. Modern farming-systems work recognizes this complexity, and includes the farmer in the adaptative processes necessary to ensure that the technology is transferable. However, most dissemination carried out world-wide is still traditionally oriented, using the less-effective demonstration techniques under scientist-controlled conditions.

Table 1 suggests the types of dissemination approach, and the components of these, commonly employed. There exist different degrees of formality at each level, such that no single approach or component would generally account for the adoption achieved, e.g. a formal Government program based on demonstration plots will soon be confounded by discussions between farmers and the informal interchange between them of seed. This means that success in dissemination cannot necessarily be ascribed to the mechanism used by the institution or the researcher to reach the farming community.

Waugh (1982) notes that the activities of different groups involved in dissemination must be coordinated. This requires that responsibilities and objectives in the program be clearly understood, and that each group or agency must support the others. He points out that adoption will not be successful if the seed promoted by extension is not available at the time the farmer needs it. There is a danger that demonstration events based on materials not yet in the multiplication stage will discourage farmers, and they may well have forgotten about the new variety by the time it appears on the market two or three years hence.

Table 1. Dissemination approaches and components

Approaches	Components
Formal programs through extension services	Demonstration plots
Formal delivery through community/farm organizations	Field days
Formal delivery through private sector contracts	Extension agents
Informal delivery by breeder, including during FSR process	Agreements or contracts
Farmer exchange	Revolving funds
	On-farm trials
	Minikits

On-farm trials overcome a lot of these problems, placing the new material directly in the farmers' hands. The availability of minikits at a field day stimulate the farmers' interest in the new variety, as he or she will almost certainly plant the material with current varieties for comparison. Such mechanisms also overcome the financial constraints small farmers have: as Martinez (1982) suggests, why should farmers pay for seed when they can produce their own? why should they spend money on a variety they do not know? and why make a trip to town, which would be an additional expense?

There is certainly a major question about the efficiency of extension services in the technology dissemination process that is particularly important when it comes to small farmers and seed. Typically, extension agents are not trained to deal with the input-variable mixed-systems of the small farmer. Most extension agents are trained in the T and V approach, which has, to date, pushed simplified technology based on fixed inputs. The dynamics of small-farm agriculture, which is subject to major risks and constraints, has resulted in the farmers developing many risk-aversion strategies, including varietal mixtures. Small farmers will evaluate a new variety themselves in this way, and, obviously, if the new variety competes well, and survives the environmental stresses placed upon it, it will be a significant component of the farmers' harvest. Invariably, lower levels of inputs are used by the farmer than extensionists recommend, rendering the evaluation process even more rigorous.

How can the extension process be improved to make it more effective in dissemination under these conditions? There clearly has to be implicit recognition of small-farm production strategies. One possible approach is to train some of the farmers themselves as part-time extension agents. This has been tried, and found to be successful (Martinez, 1982) in Guatemala. Certainly, extension agents must be made to be aware of social and community issues which influence farmers' interest in new technology. They should have the conviction themselves that there is added benefit in adopting what they are recommending. It is questionable whether this will always come about through centralized classroom-training programs.

Many agencies have attempted to establish revolving funds as a component of a seed multiplication and dissemination program. Invariably, seed production costs are higher than originally expected when this is conducted by the public-sector, and marketing problems may add to the financial burden if seed quality is not high. A revolving fund needs high standards of management if it is to be sustainable, with effective means of generating sufficient income to cover its own costs.

5. Some experiences

This workshop intends to review your experiences, and to attempt to draw out the important lessons relating to your successes in dissemination. However, there are many projects world-wide dealing with similar attempts either in plant breeding, or in the general area of increasing agricultural productivity, though in many countries public-sector seed production activities are restricted to the provision of material for commercial-scale crops. Only a small proportion of projects deal with the marginal crops grown by resource-poor farmers. The experiences of these projects are varied, but where institutional approaches to managing the seed production and dissemination process are described, it is clear that there are some valuable lessons to be learned. Some innovative projects are in fairly early stages, and the results are not yet available. Unfortunately, few projects detail the amounts of seed that pass through different channels.

1. Henderson and Singh (1990) describe efforts by the Government in the Gambia to provide seed to farmers. The principal approach was to establish, in 1972, a Seed Multiplication Unit to provide the nucleus of a seed industry. Various donor assistance was obtained for the different parts of the program. High multiplication costs resulted in a change in policy, such that the unit became responsible for seed testing and certification. It was also supposed to act as a distribution outlet for foundation seed to private contractors for multiplication. However, the responsible research units have not provided the general volumes of seed required for the unit to function properly in this way. A seed revolving fund set up to facilitate

the purchase of seed by the unit's contractors declined in value by almost two-thirds in the first two years due to bad debts.

NGOs have become important in the Gambia in seed production and dissemination. They now act as the main contracting agents for multiplying up seed each year. Training is provided by the unit to NGO personnel. A particular advantage to working with NGOs is their geographical spread into areas which Govt services find difficult to reach. The NGOs regularly meet with the unit to review progress on seed production activities. In some cases, NGOs have been selecting high-performing types from individual farms in order to multiply them up, and make them available to farms or villages with similar conditions.

2. The same authors describe work ongoing in Ethiopia, where an NGO is attempting to establish a model for local seed enterprises. The characteristics of the current local seed system are described as:

    1. Recurrent shortages of seed at the household level.
    2. Most seed transactions take place between neighbours, or through purchase at local markets.
    3. The price of seeds at the time of planting can be as high as 30-100% more than the grain price at harvest.
    4. Farmers generally cultivate 4-5 varieties of each of the main crops.
    5. Farmers practice seed replacement after about 5 years.
    6. Socio-economic interactions in the community do not necessarily allow seed borrowers to shop around on the basis of field performance of the standing crops.

    As a result of this appraisal, the NGO is attempting to develop a model based on the Ethiopian Service Cooperative, which is currently the only local operational and development entity.

3. In Nepal, a new strategy is being applied to overcome the problems of limited and uncertain seed supply, lack of adapted materials, high transportation costs, and low quality seed for the farmers (Rana and Bal, 1982). The plan includes developing a seed multiplication system in the hills, with the farmers being encouraged to produce seed for local distribution. At each hill site, a small seed house facility for processing and storing 40 to 50t of seed is being established; outlets for seed and fertilizer are also being established at strategic points to aid the flow of inputs; hill farmers are being trained in the production of quality seeds; a credit program and the extension service support the activities associated with the local production, storage and processing of seed.

    Under Nepalese conditions, it was found that any effective strategy would have to take into account several factors, many of them unique. Some are: the use of porters and mules for transportation; solar energy as the only source available for seed drying; lack of awareness of seed quality; lack of land for seed production in food-deficit areas; difficulty of encouraging agronomists and extensionists to live in remote areas; current cooperatives not in a position to play a leadership role in seed development.

4. In Guatemala, the inability of the existing seed industry to meet the high level of demand from small farmers, led the National Crop Services Agency to develop small-scale seed production and distribution among resource-poor farmers (Ortiz, 1989, reported in Ortiz, et al., 1990). Extension agents

appear to have been the main agent in this process, supporting the farmers in their seed production activities. Table 2 indicates their production in 1988.

Table 2. Small-scale seed production and distribution by resource-poor farmers in Guatemala, 1988

Crop of	No of seed plots	Production area (ha)	Seed produced (t)	No of farmers receiving seed
Maize	194	19.5	23.5	6227
Beans	153	23.6	29.5	7406
Wheat	204	13.2	29.5	3595
Potatoes	161	7.1	126.9	3635
Faba beans	6	0.2	0.4	166
Rice	1	0.7	1.4	16
Totals	719			21045

Source: Ortiz et al., 1990.

Linkages between the public and private sector in the seed industry differ from country to country. There is a general consensus that where there is heavy public-sector investment, the private sector will be discouraged. Marginal crops, of course, pose a special risk to the private sector, due to marked differences in demand from year to year. Unsold seed represents a high risk for small commercial firms.

Guatemala appears to have developed strong links between the public and private sector in the seed industry (Velasquez, 1982). In order to encourage private industry, ICTA, the national agency responsible for crop improvement, produces basic seed of most crops and offers its drying and processing facilities as a paid service to the small seed industry. ICTA also produces, processes and distributes relatively small quantities of seed in an effort to establish a quality standard and a guideline for contracting and selling prices. The strategy includes: contract seed production with carefully selected farmers at a favourable price to the producer; training of the contract producers in seed production; encouraging the producer to seek his own marketing channels rather than selling back to ICTA; provision of basic seed by ICTA, and the provision of drying, processing and bagging services for the qualified producer who wished to sell his own seed; increasing retail prices to increase the margin between the production price and the retail price of the final product.

In some cases, an excessive number of institutions appear to become involved in seed production schemes, such as examples from Alvarez (1986) and Garcia (1986). While neither author sees this as a constraint, there is a hidden cost in such top-down involvement which must, at some point, impact on the viability of the process. While mainly intended for small-farm beneficiaries in each case, the thousands of hectares planted and thousands of tonnes harvested clearly imply that the farmer is responding to institutional targets rather than to community needs.

A more producer-driven approach, where farmers participate in the definition of both the constraints and potential solutions (e.g. Gomez, 1986), suggests longer-term viability through at least partial ownership by the beneficiaries of the ideas and effort applied. The latter example is one where, with some security, one can say that it is possible to establish community-level seed businesses, managed by one or several farmers. Such businesses generally require technical assistance during their development, and are very dependent on flexible credit sources. Small farmers need seed at low-cost, a clear signal that only low-cost seed production systems will be sustainable without continuous external support.

Donor agencies have had varied results in supporting seed production programs. Successful programs were generally characterized by good management, prior experience with seed, and a good demand for seed (e.g. the World Bank's Tarai Project in India). The converse, of course, is that new seed projects in areas where there is no experience, and no initial demand for seed, will struggle for success. The Bank notes (Brown, 1982) that success is more easily achieved with relatively flexible and dynamic management than under a government or quasi-government agency in which autonomy is restricted.

IDB experience in Latin America in general, apart from observations on the inadequacy of most programs in targeting the small farmer, and in using innovative techniques, also shows the following (Ampuero, 1982):

1. With respect to seed organizations, there are few distribution mechanisms for reaching distant areas, seed quality suffers in storage and during transportation, and inflexible and rigid seed regulations reduce the amounts of seed available.

2. With respect to seed policy, governments generally do not provide incentives to stimulate seed production and establish the seed industry, policy does not clearly distinguish the roles of public and private organizations in seed production and distribution, and there are excessive bureaucratic controls in seed quality programs.

The IDB also notes that many times regulations from developed and industrialized countries have been adopted. These are often difficult to meet, and may impede the production and supply of seed to farmers. Poey (1986), in a review of some donor experiences notes that many agencies show a preoccupation with maintaining seed prices low through subsidies in order to help small farmers, though rarely is this end result achieved.

Out of these examples come some general points:

1. Some government programs are not sustainable, especially where budgetary restrictions occur and suitable trained staff cannot be retained.
2. Agencies that operate informally at the local level provide a means to support the channeling of quality seed to small farmers, and may even act as contractors in seed multiplication.
3. Seed dissemination programs for marginal agricultural populations require a greater level of effort than those targeted at more commercial operations.
4. The characteristics of seed programs, including any legislative component, should be tailored to the intended beneficiaries. Any constraints of the latter should be noted, before inflexible systems have a chance of becoming established.
5. Few programs have targeted the empresarial spirit of the small-farmer, and searched for ways to support the development of local, or community-level, seed businesses.

6. Legislation

Legislation covering seed production and dissemination varies widely. Much of it appears to be targeted at controlling this process, to ensure that genetic standards are maintained, and that seed sold to the farmer is of good quality. Some of the examples quoted above, however, suggest that legislation can act against an efficient, and, perhaps for the small farmer, an appropriate, seed industry. Certainly some authors (Douglas, 1980; Garay, et al., 1990) suggest that legislation should be the last step in the development of integrated seed programs, precisely because the controls that legislation introduces operate against efficiency and entrepreneurism. Rather than control, it is suggested that agencies responsible for seed certification should act in technical assistance roles, and that legislation should only be effected when the seed industry is operational. This is a marked contrast to the approach the World Bank has taken in most of its large seed projects.

Equally, the issue of plant breeders' rights is one that is not of equal priority throughout the developing countries. In some, PBRs do not exist, in others they are part of the legislative package at the institutional level. Few individual breeders in the LDCs would consider that effective PBRs exist, or that they benefit in any way from them.

The Technical Advisory Committee to the IARCs believes that PBRs should only be introduced after the seed industry is well along the development path (Anon, (1986), IARC position paper). It is concerned that there is ample scope for misappropriation of material emanating from the IARCs, but believes that a degree of control can be assured through provision of seed samples and varietal descriptions to certification agencies. The TAC believes that the introduction of PBRs should be left up to individual governments, and would not specifically make any recommendations in their favour.

7. Final remarks

This paper has reviewed in general terms, the principal issues that relate to seed production and dissemination, and, ultimately, adoption. While there has been a large number of seed projects, and most countries have seed programs, the actual successes of these, as they affect the small farmer, appear relatively limited, and the processes by which seed reaches the farming community are not well documented. The papers to be presented at this workshop offer an opportunity to examine this latter aspect in detail, and I hope that we will be able, as a result, to determine those strategies which have been particularly successful. There exists an opportunity to increase the impact of future breeding programs by elucidating effective dissemination mechanisms. The resource-poor small-farmers are particularly at risk from ineffective dissemination and extension methods, and we have a small opportunity to show how such methods might be improved.

8. References

Alvarez U, O., (1986). Experiencia regional de producción de semilla certificada para pequeños agricultores en Chuquisaca. In: Semilla mejorada para el pequeño productor. Segunda reunión, CIAT, Cali, Colombia, 1986. 289 p.

Anon, (1986). IARC's role in seed research and seed sector improvement. A position paper prepared for discussion at the Center Directors' Meeting, Washington, October 1986.

Anon (1987). Cereal seed industry in Asia and the Pacific. Symposium on the Cereal Seed Industry, Jakarta, 1987. Asian Productivity Organization, Tokyo, 422 p.

Ampuero, C., (1982). Support of seed projects by the Interamerican Development Bank. In: Improved Seed for the Small Farmer; Conference Proceedings, CIAT, Cali, Colombia, 1982, 179 p.

Brown, M.L. (1982). Some World Bank experience in financing seed projects. In: Improved Seed for the Small Farmer; Conference Proceedings, CIAT, Cali, Colombia, 1982, 179 p.

Cock, J. 1986. Investigación a nivel de finca y producción de semilla para pequeños agricultores: caso de la yuca. In: Semilla mejorada para el peque:o productor. Segunda reunión, CIAT, Cali, Colombia, 1986. 289 p.

Dardon, R. Ortiz, (1982). Acceptance of improved seed by the small farmer. In: Improved Seed for the Small Farmer; Conference Proceedings, CIAT, Cali, Colombia, 1982, 179 p.

Douglas, J. E. (1980). Successful Seed Programs: A Planning and Management Guide. IADS development-oriented literature series, 302 p.

Fairlie, T., Quiroz, R. and Claverías, R., 1990. Private communication. Proyecto de Investigaci<n de Sistemas Agropecuarias Andinos, Puno, Peru.

Garay, A., P. Pattie, J. Landivar, and J. Rosales, (1990). Setting a seed industry in motion: A nonconventional, successful approach in a developing country. Working Document no. 57, CIAT, Cali, Colombia, 76 p.

García, J.C., (1986). Producción de semilla de frijol de buena calidad para pequeños agricultores en el estado de Zacatecas, Mexico. In: Semilla mejorada para el pequeño productor. Segunda reunión, CIAT, Cali, Colombia, 1986. 289 p.

Gomez, F. (1986). Programa Colombiano de producción de semillas para el pequeño agricultor. In: Semilla mejorada para el peque:o productor. Segunda reunión, CIAT, Cali, Colombia, 1986. 289 p.

Henderson, P.A. and R. Singh, (1990). NGO-Government links in seed production: Case studies from The Gambia and Ethiopia. . Network Paper 14 in Agricultural Administration (Research and Extension) Network, ODI, June 1990.

Martinez, E. (1982). Successful extension methods for introducing new varieties and increasing the use of seed. In: Improved Seed for the Small Farmer; Conference Proceedings, CIAT, Cali, Colombia, 1982, 179 p.

Ortiz, R., Ruano, S., Jurez, H., Olivet, F. and Meneses, A. 1990. Closing the gap between research and resource-poor farmers: A new model for technology transfer developed in Guatemala. OFCOR - Practical Experience Paper, ISNAR, 1990.

Rana, P.N. and S.S. Bal, (1982). Experience with small farmers in the use of improved seed in the hills of Nepal. In: Improved Seed for the Small Farmer; Conference Proceedings, CIAT, Cali, Colombia, 1982, 179 p.

Velasquez, R. (1982). The seed business in the transfer of technology to farmers: The Guatemala experience. In: Improved Seed for the Small Farmer; Conference Proceedings, CIAT, Cali, Colombia, 1982, 179 p.

Waugh, R.K. (1982). Seed in the Transfer of Technology to Small Farmers. In: Improved Seed for the Small Farmer; Conference Proceedings, CIAT, Cali, Colombia, 1982, 179 p.

Wooley, J. 1986. Investigación a nivel de finca y producción de semilla para pequeños agricultores: caso de frijol. In: Semilla mejorada para el pequeño productor. Segunda reunión, CIAT, Cali, Colombia, 1986. 289 p.

ARTISANAL SEED SUPPLY SCHEMES: A STRATEGY TO EXTEND THE DEVELOPMENT OF ORGANIZED SEED SUPPLY SYSTEMS TO MEDIUM AND SMALL FARMERS

Adriel E. Garay*

PROBLEM DEFINITION

Highly productive seeds are needed to support agricultural growth, production, and productivity at the large-, medium- and small-farmer level. In Latin America, while the utilization of improved seeds** has advanced in commercial/industrial crops, the estimated rate of use is below 40% in rice and 10% in beans, 1% in cassava, 10% in tropical pastures, and 10% in open-pollinated maize (seed unit estimates). It is generally accepted that almost 75% of farmers in the tropical world have not yet benefited from modern technologies being generated by research (CIAT, 1982). This problem is accentuated in medium- and small-farmer production systems. Despite this low rate of use of improved seeds, the contribution of small farmers to food production in Latin America is estimated at 32% for rice, 77% for beans, 51% for maize, and 90% for cassava (CEPAL, 1982)***, indicating that they have a very important role to play in the future. Their contributions in specific regions are much higher.

Several seed supply schemes coexist within countries: corporate, conventional and traditional. The corporate schemes can be characterized as large, conglomerate organizations of national or multinational scope, carrying out research/production/marketing functions, present in large, uniform, prime markets such as those for hybrids and commercial/industrial crops. This system has proved to be effective in responding to market changes, financially sustaining its overall activities, and assuring the quality of its product and services. Consequently, it has acquired prestige, visibility, and competitiveness. Despite all its strengths, this system has not delivered to small, scattered diverse, and risky (SSDR) markets such as those found under medium- and small-farmer conditions in the developing world.

Another scheme that most national and international development programs favored in the past is the conventional scheme. These are typically indigenous schemes based on public support services in terms of varietal development, but with production/marketing operations in the hands of public and private organizations. The conventional scheme also operates in large, uniform, and low-risk markets. Like the corporate scheme, it is capital intensive, requiring large single jumps in investment. It has also shown that it can supply seeds of assured quality and can be financially sustainable when privately operated. Even though effective under large, commercial farming systems, this scheme has not delivered improved seeds to SSDR markets either. Since large seed companies have centralized production units to take advantage of economies of scale, the overall process of transfer and adoption is complicated, and small communities cannot be reached in an effective fashion.

* Senior Seed Scientist, Seed Unit, Centro Internacional de Agricultura Tropical (CIAT), Apartado Aéreo 6713, Cali, Columbia. Paper to be presented at the Seed Production Mechanisms Workshop, Singapore, November 5-9, 1990.

** Improved seed, in its modern context, is a biological technology requiring physical quality attributes (viability, health, vigor, purity, etc.) to be an effective carrier of biogenetic innovations from the research phase to the farmer's fields.

*CEPAL and FAO, in López Cordovez, 1982

In some cases, government corporations intervened in production and marketing of seeds. They sometimes attempted to reach farmers with subsidized seeds. However, this socially well-motivated approach has not been sustainable and enduring (Garay et al., 1989). In the meantime, a large segment of farmers in the developing world do not use improved seeds.

In contrast to the above two organized schemes, the major source of seed in the developing world is by far the traditional seed supply (Grossman et al., 1988). This scheme can be typified by the farmer saving his own seeds, or obtaining them from his neighbors or in the local grain market. This scheme has several positive features: the farmer can do the work, the seed is available where farmer investment required is minimal, and the farmer has good knowledge of his seed's potential. A close analysis indicates that this system should not be interpreted as static; on the contrary, it is a dynamic system lending itself to farmer-to-farmer flow of seeds and it can produce rich dividends, if linked to modern technologies generated by production methods. Consequently, the replacement of obsolete varieties is very slow (Rajbhandary and Bal, 1989) and maintenance of physical quality of existing verieties is erratic.

From this brief analysis, it is evident that development of seed supply cannot be treated monolithicly under a single strategy; instead, different schemes are needed to effectively deliver improved seeds to different segments in the market. The weaknesses of existing schemes vis-a-vis the need for improved seeds in SSDR markets clearly indicate that development strategies could ill-afford not to develop relevant alternatives.

Therefore, purposeful and clearly focused research and development strategies are needed to develop seed supply under medium and small farmers. When entering SSDR production systems where medium and small farmers predominate , most classical textbook seed technologies do not seem relevant. However, a skillful combination of lessons learned in modern seed schemes with the best features of traditional schemes, creates a new and ample opportunity for development. A new scheme of intermediate nature between the conventional and traditional schemes would gradually bridge the existing gap between modern and traditional seed supply. This evolving scheme is actually a family of new approaches currently beginning under the names of nonconventional, artisanal, on-farm, participatory, community, and local schemes.

THE ARTISANAL CONCEPT, METHODS AND RESULTS

Recent research and empiric observations indicate that the solution to SSDR systems may be in the development of simple, relevant, and not too costly seed supply schemes. This proposition that may not have been conceivable in the past, now seems feasible due to new evolutions that facilitate the process. For example, advances were achieved in improvement of varieties, many services were created to support seed and crop production, and seed production technologies are advancing.

Farmer-producer organizations (FPO'S) and a range of government and nongovernment organizations (NGOs) interested in production technologies have also evolved. All these and other factors create ample opportunities to expand seed supply systems beyond large commercial operations.

Several new cases demonstrate the feasibility of artisanal menthods (Ortiz and Trejos, 1988; PROGETTAPS Report, 1989; Garay et al., 1989; CIAT, 1987;

Rajbhandary and Bal, 1989). To illustrate some methods and achievements, some evolving cases in Latin America will be briefly described.

Colombia is a country with advanced organized seed supply schemes. Corporate and conventional seed schemes are very dynamic in hybrid sorghum, rice, and soybeans. With the exception of a small (20,000kg) supply of an old variety by a government organization, the bean seed supply is practically nonexistent. In 1983, a small cooperative (COAGROSANGIL) initiated artisanal seed production. The project that had started in 1983 reached 30,000kg in 1986, becoming the largest seed supplier in the country. Since then, innovations on several fronts (production methods, marketing network, incorporation of new varieties, and expansion of production volume) are being advanced by the cooperative. This pilot case is generating interest on the part of other FPOs and government officials. Recently, another small cooperative that had been active in participatory research (ASHORTOP, Pescador, Cauca) has successfully initiated similar attempts. In the first attempt, 2,000kg of seed were produced, which will increase to 5,000 - 10,000kg in the second year. Similar schemes are being started in cassava seed production, based in local, organized cassava-drying cooperatives.

In the 1960's, the organized seed supply in Guatemala was in the hands of the government. In the 1970's the strategy changed in order to promote the evolution of private suppliers, based on government support through basic seeds, quality control (certification), credits, etc. A very dynamic supply of hybrid maize evolved. But bean seed supply and highland open-pollinated maize were practically unaffected. In the late 1960's a technology transfer project (PROGETAPPS) incorporated artisanal seed production as a central strategy. New varieties of beans were rapidly produced by local farmers and passed on to their neighbours through various sale/exchange/share arrangements. In the first pilot region (Jutiapa), production in the first year reached 2,727kg, which is increasing at a rate of 10,000kg a year. The scheme has been gradually expanded to the whole country and to other crops. Realizing that just extending varieties will not generate a long-lasting supply system, Guatemalans are now incorporating the enterprise development concept with organized FPOs. In the new scheme, field production, post-harvest processing and marketing will be carried out by the FPOs, while government agencies would offer assistance to them, providing basic seeds and technical assistance to promote quality seed production and market development.

Among Latin American countries, Bolivia has started to develop organized seed supply systems most recently. In organizing their system, different strategies were used under large-, medium-, and small-farmer conditions. The main features in adjusting strategies to medium and small farmers were: creating local seed organizations; simplifying production methods to allow entry to the process; and using a seed certification service in promoting quality seed production rather than policing. As a result, 55 participatory, nonconventional, dynamic, and production/marketing enterprises have developed. Among these, half can be characterized as medium to small enterprises, producing seeds with hundreds of small farmers.

A brief diagnosis across countries that are beginning small-scale seed production indicates that the lack of simple methods and tools for post-harvest management of seeds is a serious constraint. Aware of this need, CIAT has started research on production technologies. In beans, inexpensive but highly effective methods and tools are being achieved to facilitate harvesting, drying, cleaning, and quality assessment (Camargo et al., 1989; Garay et al., n.d.). In cassava, effective methods based on good field agronomy, selection at harvest, and preparation and treatment of stakes are being put together. Results achieved with beans have triggered the initiation of new pilot projects by national programs in Honduras, Panama, Ecuador, and Peru. Research on rice, maize, and pasture seeds is beginning as well.

LESSONS BEING LEARNED

Field results in this area of research and development indicate that the development of seed supply systems under medium- and small-farmer conditions is feasible. There is increasing interest on the part of national government, nongovernment, and farmer-producer organizations in this approach, indicating opportunities for development beyond the magnitude of pilot projects. Since each case has different needs, it is probably advisable to avoid models. However, some of the useful lessons identified when looking across cases that are showing success follow.

1. **Do not confuse extending varieties with developing seed supply systems.** Realizing that having good varieties at the gate of research institutions is not enough, some research and extension programs have gone one step further by giving extension services the responsibility to extend the varieties. Some technology transfer projects focus on procuring improved seeds to distribute among medium and small farmers. These methods fail to recognize the innovative ability of these farmers. Other projects provide small quantities of improved varieties and assist farmers in multiplying in their community, to facilitate farmer-to-farmer dissemination of seeds. Though effective in introducing varieties, even in the best of cases these approaches, which extend only the variety, quite frequently do not survive beyond the life of the project. Production and marketing of seeds need to be instituted in the form of commercial enterprises even if only on a small scale so that seeds can be delivered in a continuous and financially sustainable fashion.

2. **Build on organized farmers.** It is economically risky to think that new organizations have to be developed for the sole purpose of producing and marketing seeds. On the other hand, there is an abundance of organized farmer-producers (FPOs) or individuals who already have ongoing activities with some economic base (Camargo et al., 1989). These organizations may be cooperatives, committees, associations, etc., already dealing with supplying inputs, marketing produce, and channeling credit, among other activities. Experiences in Bolivia, Guatemala, Colombia, and Panama create a serious suspicion that existing organizations have a potential to develop a built-in capacity to produce and market seeds and so far they are under utilized. Their organization makes channeling information, credit, and technical assistance easier, thereby being a good multiplying factor for the dissemination of improved seeds and related technologies. This creates the opportunity to add seed supply as a new line of product or service readily recognized by constituents. In the process, these production units become key links in a chain that joins research with farmers' fields.

3. **Start small and simple.** Many seed development projects in the past failed due to subjective and overenthusiastic estimation of the market. In Latin America, it is common to find large seed-conditioning facilities that hardly utilize more than 10% of their capacity. The same mistake can ill be afforded in the SSDR markets. Instead, these situations seem to lend

themselves to smallness as a condition for beginning a process. Smallness and simplicity, however, should not be confused with deficiency or mediocrity. Scientifically sound principles and methods are needed to assure the delivery of quality seeds at low cost. The small initial pilot units permit adjusting strategies and methods without great risks of failure. Once enough experience has been gathered, more complex methods can be incorporated if needed and the operation can be enlarged following changes in the market. This has been the case in beans and wheat in Bolivia, rice in Daule, Ecuador, and beans in San Gil, Colombia, and Jutiapa, Guatemala.

Past and current experiences demonstrate that if the process is allowed to start, even with a small-artisanal nature in the beginning, it will evolve and become more specialized with time and experience, if given room to operate. A high level of specialization should not be a requirement to begin. This makes simple, local, artisanal seed supply schemes an attractive approach to extend both supply and utilization of improved seeds under medium- and small-farmer conditions.

4. **Assure availability when needed and where needed.** Seeds supplied to farmers should have better quality than the seeds saved by the farmers themselves (Delouche, 1982). Some orthodox seed developers would prefer perfect-quality seeds from the outset. However, availability of reasonably good seeds when and where needed seems to be more important that nonexistent or scarce perfect seeds.

   Sophistication in technologies aimed at perfection in quality to the point where it is no longer affordable by the majority of potential participants may be limiting development. The perfection objective led to the establishment of hard-to-achieve norms and procedures, which in the long run inhibited participation in the system. In contrast, the approach that seems to allow participation in and initation of the process seems to be flexibility, focusing on availability. Quality should be one of those features that is good enough to start with and then perfect over time.

5. **Differentiate the product.** It should be recognized that in non-hybrids, such as open-pollinated varieties and clonally propagated crops, all farmers are virtually seed producers. In theory, once they have access to a new variety, they can keep seeds for subsequent plantings year after year. However, recent evidence is showing that the lack of abilities and environmental stress create the need for dependable sources. And a market is developed gradually as a result of specialized supply and awareness of the advantages of improved seeds by farmers. To take advantage of this phenomenon, a seed enterpriser needs to differentiate his seeds from common grain regardless of the size of the operation. This has been universally used in corporate and conventional seed schemes with highly successful results in the past. One of the simplest ways to differentiate improved seeds has been distributing them in bags that clearly show brand name, type of seed, basic quality features (purity, germination), weight, etc.

   This information can be printed on the bag or attached as a tag. It is being recognized that even if seeds are not certified, this information is extremely valuable in gaining visibility for good suppliers, repeating sales, expanding the market, and competing through quality and price. Depending on the development stage and sophistication level of the consumer, this differentiation can be very simple or elaborate.

6. **Don't control prices.** Paternalistic schemes such as seeds at subsidized prices and market interventions controlling artificially low prices in the best of cases have given short-lasting results. There is a clearer understanding that improved seeds are technologies that need to be produced and sold. Somebody needs to develop a special capacity to produce them and make a business of it so that improved seeds can be supplied in a continuous and growing fashion. In the past, trying to control prices has been a frequent error that inhibits investment in seed production and marketing, which in turn blocks the transfer of this productive technology to farmers' fields. One loser in the process is the farmer, who will not profit from more efficient seeds. Other losers naturally are the final consumers due to insufficient production and increased prices.

   Corporate and conventional seed systems demonstrated that to develop a seed industry on a sound economic basis, prices must be defined by market forces. Without this, the competitive aspect of the market and the incentive to innovate are lost, and financially sustainable seed supply systems cannot be developed. This principle is even truer when promoting investment in seed systems to supply to medium and small farmers.

Some lessons are being learned. One clear lesson is that even in the most remote and apparently "resource-scarce situations," it is possible to develop seed supply systems if rigid conventionalisms in the approach are overcome. Much ground remains to be covered. There is a need for research and development. Research needs to be development oriented and with easy implementation in mind. Both biological production methods as well as social-organizational technologies are needed to incorporate the farmer as the central actor in the process.

There is growing evidence that corporate, conventional, and artisanal schemes have a role to play under different market situations. Special efforts in terms of policies, strategies, and specific actions are needed to facilitate their development. Most countries are interested in principle in the artisanal scheme, but potential groups need to be identified, trained, and financed. National research and development organizations need to provide key services, and some classic barriers need to be overcome. In summary, development projects with a clear objective of stimulating this scheme will be needed. Finally, by supporting the development of seed supply for medium and small farmers, a greater return to investment in crop research, agricultural growth, equity, and food security will have been furthered.

BIBLIOGRAPHY

CAMARGO, C.P.; Bragantini, C.; Monares, A. 1989. Seed production systems for small farmers: A nonconventional perspective. CIAT, Cali, Colombia.

________ et al. 1989. Seed for small farmers: Support infrastructure. CIAT, Cali, Colombia.

CIAT. 1982. Improved seed for the small farmer: Conference Proceedings. CIAT, Cali, Colombia.

________. 1986. Produccion de semillas mejoradas para pequenos agricultores: Memorias de un taller de trabajo. CIAT, Cali, Colombia.

________. 1987. Small-farmer cooperative produces improved bean seed. CIAT International 6(1).

DELOUCHE, J.C. 1982. Seed quality guidelines for the small farmer. In: Improved seed for the small farmer: Conference proceedings. CIAT, Cali, Colombia.

GARAY, A.E.; Pattie, P.; Landivar, J.; Rosales, J. 1989. Setting a seed industry in motion: A nonconventional, successful approach in a developing country. Working document no. 57, CIAT, Cali, Colombia.

________; Aguirre, R; Giraldo. G. n.d. Tecnologias artesanales para el manejo poscosecha de semillas. CIAT, Colombia. (In press.)

GROOSMAN, T.; Linnemann, A.; Wierema, J. 1988. Technology development aand changing seed supply systems. Research report no. 27, Development Research Institute, The Hague, Netherlands.

ORTIZ, R. and Trejos, J.A. 1988. Semilla mejorada para pequenos agricultores de Guatemala. Boletin de Semillas CIAT 8(2).

RAJBHANDARY, K.L. and Bal S.S. 1989. Private (small scale) producer-sellers seed program: An innovation for seed dissemination in the hills of Nepal. Progress Report.

WELLARD, K.; Davies, P. 1990. The state, voluntary agencies and agricultural technology in marginal areas. Agricultural Administration Network Paper 15. ODI, London, England

SEED PRODUCTION AND DISTRIBUTION MECHANISMS

Case study of pigeonpea in Kenya

P.M. Kamani[1] and O.L.E. Mbatia[2]

[1]Department of Crop Science and [2]Agricultural Economics
University of Nairobi
P.O. Box 29053 Nairobi, Kenya

ABSTRACT

Five hundred pigeonpea farmers in Machakos, Embu and Kitui districts were interviewed in September and October, 1989 to determine the adoption of the improved early maturing pigeonpeas, the area planted, and to evaluate the alternative seed production and distribution mechanisms used by small scale farmers. Seventy five, 36.6 and 72.6 percent of the farmers in Machakos, Kitui and Embu, respectively were growing the improved short duration pigeonpea cultivar, NPP 670. About 51 percent of farmers first heard about the new cultivars through agricultural extension officers, 22.9 percent from their neighbours, 8.1% saw them on neighbours field, 3.6% in farmers training centres and 0.4% through radio broadcast. Seventy nine percent of the farmers practised intercropping. Average cropped land was 0.63 hectares. Maize occupied 20.7 percent of the cropped area, pigeonpea 19.6 percent, beans 18.7 percent and cowpeas 14.5 percent. Forty percent of the farmers first obtained seeds of improved pigeonpeas from agricultural officers, 31 percent from neighbours, 11 percent direct from pigeonpea project, 9.4 percent from local shops and 6.7 percent from local markets. In 1989, 37.3 percent of farmers planted their own seed, 12.5, 6.6 and 1.7 percent bought seeds from agricultural officers, neighbours and local markets, respectively. Pigeonpea seeds are produced by small scale farmers on contract, pilot seed multiplication project, Machakos Integrated Development Project (MIDP), pigeonpea project and to a lesser extent women groups. Farmers indicated that pests were the main constraint in growing pigeonpeas. Most farmers obtained as much seed as they required. Local seed companies do not multiply pigeonpea seed due to fluctuations in demand for seed. It is suggested that the Ministry of Agriculture contract seed companies and farmers to produce and distribute pigeonpea seeds. Organised marketing of pigeonpeas is required.

INTRODUCTION

Pigeonpea (Cajanus cajan (L.) Millsp) is one of the most important legume crops in Kenya because it is drought tolerant and is a multi-purpose crop. Its seeds are a major source of protein (17-28%) for resource-poor families, stems are used as fencing material and fuelwood, leaves, pods and damaged seeds as animal feed (Kimani, 1985). It is grown in semi-arid areas which have unreliable rainfall of less than 800 mm annually. The major growing areas in Kenya are in the Eastern region comprising Machakos, Kitui and Embu districts, and to a lesser extent in Central, Rift Valley and Coast Provinces. It is however found in retail markets and shops throughout the country. An estimated 115,000 hectares are under pigeonpea in Kenya. Kenya is the world's second largest producer of pigeonpea after India (Onim, 1981).

The yield of pigeonpea is low in Eastern Africa averaging 450 to 670 kg ha[1]. Research workers have projected that yield could be improved up to 1120 kg ha[1]. It has been shown that under research conditions, a yield of between 2637 to 4250 kg ha[1] can be realized (Onim, 1983; Kimani, 1988). In Australia, yields as high as 7500 kg ha[1] have been recorded under research conditions (Akinola and Whiteman, 1972).

The factors contributing to low yield in Kenya are inferior varieties, diseases and pests, moisture stress and drought, poor soil fertility and poor crop husbandry practices (Kimani, 1987). The socio-economic factors are poor prices, poor marketing and infrastructure (Mbatia and Kimani, 1987).

Research aimed at improving the production of pigeonpea started in 1976 at the Department of Crop Science, University of Nairobi, Kenya, partly as a consequence of some earlier work at Makerere University, Uganda. (The pigeonpea research at the University of Uganda was fully funded by the International Development Research Centre (IDRC), Ottawa, Canada.) The major objectives of the pigeonpea project are to :

i) Develop new cultivars of pigeonpea that are high yielding and with acceptable pod and seed characteristics.

ii) Develop cultivars that have resistance to Fusarium wilt, the most important pigeonpea disease in Kenya.

iii) Develop suitable agronomic practices for the new cultivars.

iv) Initiate a seed multiplication and distribution scheme. The project also has a training component for graduate students in agronomy, plant breeding and plant pathology.

Most of the above objectives have already been achieved. Through crossing and selection, the researchers have developed and released cultivar NPP 670 which matures in four and a half to five months compared to traditional pigeonpeas which mature between 10 and 12 months (Kimani, 1987). NPP 670 has a high yielding potential and two crops are harvested before traditional pigeonpeas mature. It is moderately resistant to Fusarium wilt. Other cultivars already developed include Kitui 1, NPP 673/3, Kioko and Munaa. The latter take 6 to 7 months to mature and have excellent seed characteristics and yield potential. Major evaluation of these cultivars started in 1983. Between 1983-89, 40 on station trials and over 140 farm trials were conducted. Average yields of the former reached 2400 kg ha[1], compared with 1500 kg ha^{-1} on farm.

Although resistance to Fusarium wilt has been found and incorporated into a new generation of early maturing cultivars, insect pests remain a major obstacle to higher productivity of pigeonpeas in farmers fields (Kimani, 1989, 1988, 1987; Okiror, 1986). Research work on control of pigeonpea pests is still going on. Mbatia and Kimani (1987) carried out a social-economic survey of pigeonpea farmers in Machakos, Embu and Kitui districts. A sample of 1500 farmers were interviewed. The study showed that the major problem farmers encountered with these new varieties were insect damage on pods and seeds and diseases to a lesser extent. Farmers liked the improved varieties because of their early maturity and high yields.

SEED MULTIPLICATION AND DISTRIBUTION

In Kenya, seed multiplication and distribution is primarily conducted through the private sector. The main seed companies include Kenya Seed Company which handles the largest volume of seed trade. It multiplies and distributes seeds of cereals such as maize, wheat, barley as well as vegetable seed (tomatoes, carrots, beans) and pasture seed. It has its main offices in Kitale, Rift Valley provinces with branches in Nairobi and other urban areas. The East African Seed Company located in Nairobi multiplies and distributes vegetable seeds, fertilizers and crop protection chemicals. Oil Seeds Development Company, a subsidiary of East Africa Industries mainly handles oil seeds especially sunflower and rape seed in collaboration with Kenya Seed Company. Also there are other small companies in seed trade. The large companies contract large-scale well-established farmers to multiply seed and offer premium prices for seed crop compared to the general commercial crop. Seed certification and quality control is carried out by the National Seed Quality Control Service (NCQS) based in Lanet, Nakuru district in collaboration with seed companies. NCQS performs crop inspection and issues certificates for quality seed in conformity with international seed regulation.

In Kenya new crop cultivars are mainly developed by breeders in the public sector, although the private companies maintain research departments which include breeders. Promising crop cultivars are usually entered in the national performance trials (NPT) for three years and the best performers are recommended for release through the National Variety Release Committee (NVRC) which draws its membership from government ministries, public institutions and private companies. After multiplication, sorting, packaging, seeds of new varieties are distributed through a network of co-operative stores, private shops and quasi-government farm input stores such as the Kenya Grain Growers Co-operative Union (KGGCU) throughout the country.

The University of Nairobi had originally intended to turn over the seed of the new pigeonpea cultivars to the private sector, but the latter is more interested in more lucrative crops such as maize, wheat, barley, vegetable crop and pasture seeds. Companies found through study that demand for crops grown under semi-arid conditions, such as pigeonpeas, green grams and cowpeas, is very unstable. Since most of these are composites or open or self-pollinated cultivars, farmers produce their own seed and would only buy seed if previous year was dry. Demand was not firm.

Mechanisms of Seed Multiplication and Distribution

The pigeonpea project has pursued five multiplication mechanisms:

1. Directly by the project

Multiplication was started in 1983, land being leased in Machakos district. The project carried out land preparation, sowing, weeding, pest control, irrigation, harvesting, drying, sorting, and packaging of the seeds.

2. Machakos Integrated Development Project (MIDP)

MIDP is a rural development project funded by the European Economic Community (EEC) and the Kenya Government, and located in Machakos town. Among its objectives is to provide farm inputs and technologies to small scale-farmers in Machakos district. The University provided foundation seed and field inspection of seed crops, while MIDP did all the remaining production, packaging and distribution of seed packets.

3. **Small-scale farmers contracts**

The project chose farmers who had already grown improved pigeonpeas for at least one season. Technical assistance and chemicals were provided since most of these farmers are very poor and cannot provide any funds. Farmers provided all the labour. Farms were visited three times during the cropping season for roguing. Even in the predominantly self-pollinated pigeonpeas, there is a large amount of outcrossing (Onim, 1981; Kimani 1987; Githiri and Kimani, 1988). Farmers were paid on spot at harvest (1988, Kshs 7.00 kg^{-1}; 1989, Kshs 8.00-9.00 kg^{-1}). However, market prices rose to Kshs 12.00 kg^{-1} in 1988, and farmers demanded the project increase the price it paid.

4. **District based pilot schemes**

In order to ensure that adequate seed was produced close to the demand areas, pilot seed schemes were initiated with the support of the district agricultural officer. These were generally felt to be the basis of a self-sustaining system. Seed is multiplied in farmers training centres (FTC) and farmer's fields and purchased by the district agricultural offices, sorted, treated, packed and resold to farmers. It was initiated in 1984 in Kitui district with an initial sample of 20 kg of foundation seed and had increased to 6 t in 1988/89. The project provided foundation seed for multiplication which is renewed after every three years.

5. **Women's groups**

The project started working with women's groups in 1988/89 cropping year. Multiplication of pigeonpea seed by women's groups was co-ordinated through the women's bureau, Ministry of Culture and Social Services. In this arrangement the groups were to lease land, preparation, sowing, weeding, roguing (with project staff), harvesting and sorting. The project would provide seeds, chemicals and spraying instructions, inspection, purchase of seed, packaging and distribution.

In addition, the East African Seed Company was multiplying seed in Meru district. The company carried out all the operations including packaging, labelling and distribution of seed.

Improved pigeonpea seed is disseminated through :

1. Direct sales from the project - at Kshs 17.00 per kg. Seeds are sold either at project headquarters at Kabete, or throughsubstations at Makueni, Machakos, Embu, Kitui.

2. Agricultural extension offices - at district, division or location levels.

3. Co-operative union stores in target areas.

4. Open air markets and shops. Seed mainly originates from farmers fields.

5. Farmer to farmer.

6. Private companies especially East African Seed Company. Their operations are mainly limited to Nairobi and Meru.

The distribution of seed through these channels is mainly by informal contracts. The recommended base price is Kshs 8.50 per a 500g polythene bag in which seeds are normally packed. However, prices at sale points vary between Kshs 8.50 to Kshs 15.00.

OBJECTIVES OF THE STUDY

The general purpose of this study, was to generate and disseminate information leading to the strengthening of systems for the production and dissemination of improved pigeonpea seeds. The specific objectives were :

1. To estimate the amount of improved pigeonpea seeds distributed, the number of growers using improved seeds and the area planted by the small scale farmers.
2. To describe and evaluate the alternative seed production and distribution mechanisms being used by small scale farmers.
3. To make policy recommendation on improved pigeonpea seed production and distribution mechanisms.
4. To make the results and conclusions of the study widely known.

METHODOLOGY

Sampling technique

The researchers visited the district agricultural officers (D.A.O's) at Machakos, Embu and Kitui to establish a working relationship and explain the purpose of the survey. The D.A.O's assured the investigators that they would inform all the field officers and request them to support the investigators and recruit the enumerators. It was agreed that the study was timely in that information would be gained on how farmers acquired improved pigeonpea seeds and on help they required in seed multiplication and distribution.

Questionnaire

The questionnaire was prepared in consultation with extension workers in the field, agricultural officers and social scientists. The questionnaire was structured and had 77 primary questions. The questions were mainly on improved pigeonpeas. The major variables in the questionnaire were those related to seed production and distribution mechanisms. The variables included labour inputs, area planted, sources of improved seeds, price paid on seeds, yields, methods of harvesting, marketing channels, and other related variables.

Enumerators

The enumerators were recruited and hired from each area where the study was to be conducted. Equal numbers of male and female enumerators were selected. They were conversant with the area and spoke the local language. They were trained for a total of 8 hours. Many of the enumerators had done similar work elsewhere. All the variables in the questionnaire were translated to local vernacular. At the end of the training each enumerator was given 5 questionnaires to practise with. The researchers went through completed questionnaires with enumerators to discuss any problems which they might have encountered. A final questionnaire was drawn after pre-testing.

A total of 500 farmers was planned for interview. This included a random sample of 200 from Machakos, and 150 each from Embu and Kitui districts. In each district the area with a high concentration of farmers growing improved pigeonpeas was selected as a research area. In Machakos district, Masii and Makueni divisions were selected. It should be noted that on-farm testing of improved pigeonpea started in Makueni area and has the highest number of farmers growing the improved pigeonpeas. In Embu district, Karaba and Gachoka locations were selected. The two areas have been used for seed multiplication. Kitui Central and Mwingi divisions, in Kitui district were selected because of high concentration of farmers growing both traditional and improved pigeonpea.

The random sample of farmers to be interviewed was drawn from a list of farmers growing pigeonpeas in targeted areas. The list was provided by extension officers working in that area. The list was drawn at random. Each enumerator was given a list of farmers to be interviewed. The enumerator was assigned to interview two farmers per day. In case the farmer was not at home or unavailable for interview, a new farmer was substituted or interviewing time rescheduled.

The final interview included 472 farmers. Of these 212 were from Machakos district, 135 from Embu and 125 farmers from Kitui district. The whole interviewing exercise went on smoothly due to the good cooperation of the farmers and extension workers.

RESULTS OF EMPIRICAL ANALYSIS

The socio-economic aspects

In all three districts, 54 percent of the 472 farmers interviewed were males and 46 percent were females. During the interviewing it was found that where both husband and wife were present they consulted each other in answering the questions.

Household composition

The information on household was gathered according to age, family size, number of children going to school and employment. The average ages in Machakos, Kitui and Embu were 49, 45 and 45 years respectively. Average age in all three districts was 47 years. The family size in the three districts was eight persons per household with an average of four children going to school. Some of the children lived at home with their parents but 50 percent of the parents stated that their children either were working elsewhere or going to school away from home.

Farm labour

The small scale farmers mainly use family labour and hired labour during planting, weeding, harvesting and land preparation. About 58 percent of the farmers indicated that they hired some labour for farm work and 42 percent did not hire since the family labour was sufficient or could not afford to hire. The majority of farmers hired labour on a part-time basis. Thirteen percent of farmers hired one worker on full time and 7.4 percent hired two workers on full-time basis.

Table 1 shows that an average of two workers were hired for land preparation, 2.3 for planting, and 4 workers for weeding. Weeding required more workers than any other activity. Some farmers hired workers during harvesting time. The farmers also indicated that they also kept some livestock.

Table 1: Hired farm-labour activities, 1989.
Units: percentage of surveyed farmers

		District		
Activity	Machakos	Kitui	Embu	Mean
Land preparation	31.2	39.2	23.7	31.1
Planting	21.7	34.4	44.4	32.6
Weeding	23.6	43.2	52.6	37.1
Domestic (household)	14.2	14.4	5.2	11.7
Looking after animals	17.5	22.4	19.3	19.2
Other work	3.3	0.8	3.0	2.5

On average they kept 7 head of cattle, 10 goats, 5 sheep and 21 chicken. It could be inferred that farmers kept livestock for selling to earn income and perhaps for manure.

Land

The four major factors of production are land, labour, capital and management. Although all of these are important in production, land is perhaps the most vital for the farmer. In the three districts, the estimated land per farmer was 7.5 ha. Of this, 3.1 ha were under cultivation. The rest was used for grazing, source of woodfuel or bush. In many cases in the semi-arid area, the land has yet to be surveyed and consolidated. The average land under cultivation and owned by the farmer should be regarded as estimates since no measurements were taken by enumerators.

Education

Nearly 70 percent of the farmers had gone to school or attended some educational training. About 62 percent had a primary school or secondary school education.

Decision making on the farm

In many cases it has been assumed that decisions on cash crops to be grown are made by men whereas that on food crops for domestic consumption is made by women. Table 2 shows that 56.3 percent of the farmers indicated that both husband and wife make decisions regarding crops to be grown.

Table 2. Who made decision on what crops to be grown in 1989 cropping season

Decision maker	Machakos	District Kitui %	Embu	All
Wife	34.8	11.3	10.5	22.1
Husband	23.8	26.6	15.0	21.6
Both	41.4	62.1	74.0	56.3

CROPPING SYSTEMS

The cropping system normally practised by small scale farmers is mainly intercropping. Mbatia and Kimani (1987) found that 81.4, 83.3 and 73 percent of farmers in Machakos, Embu and Kitui respectively practised intercropping. They intercrop maize, beans and pigeonpea. During the 1988/89 season. The estimated average area under crops was 0.65 ha. In Machakos, farmers had a bigger area of 0.85 ha under crops, followed by Kitui with 0.5 ha and Embu with 0.44 ha.

Table 3 shows the average crop area for various crops grown in the three districts. Maize had the highest crop area followed by cotton.

Table 3: Area in hectares under various crops in farmers fields in Machakos, Kitui and Embu districts in 1988/89.

Crop	Machakos	Kitui	Embu	Mean
Maize	1.30 (21.0)*	0.66 (20.6)	0.73 (20.4)	0.90 (20.7)
Beans	0.70 (19.5)	0.55 (18.7)	0.44 (17.5)	0.56 (18.7)
Pigeonpeas	1.01 (20.4)	0.60 (19.4)	0.30 (18.8)	0.64 (19.6)
Cowpeas	0.52 (15.3)	0.38 (16.9)	0.31 (11.3)	0.40 (14.5)
Cassava	0.35 (3.1)	0.24 (4.4)	0.21 (0.8)	0.27 (2.8)
Cotton	1.06 (6.0)	0.54 (0.5)	0.72 (8.8)	0.77 (5.4)
Millet(Finger)	0.44 (5.7)	0.39 (7.3)	0.29 (5.8)	0.37 (6.1)
Sorghum	0.41 (5.1)	0.44 (7.0)	0.19 (3.2)	0.35 (5.0)
Fruits	0.61 (1.4)	0.14 (0.9)	0.22 (0.5)	0.32 (1.0)
Bananas	0.13 (0.5)	0.13 (1.2)	0.17 (2.1)	0.14 (1.2)
Green grams	0.46 (1.1)	0.34 (1.7)	0.37 (5.5)	0.39 (2.5)
Coffee	0.81 (0.2)	0.20 (0.2)	0.20 (0.2)	0.40 (0.2)
Potatoes	0.25 (0.6)	0.18 (0.7)	0.18 (1.1)	0.20 (0.8)
Sunflower	0.40 (0.1)	-	0.51 (0.8)	0.13 (0.3)
Wheat	0.10 (0.1)	-	-	0.17 (0)
Vegetables	0.05 (0.1)	0.04 (0.3)	0.40 (0.2)	0.16 (0.2)
Dolichos	-	-	0.38 (3.0)	0.13 (0.9)
Tobacco	-	0.40 (0.2)	-	0.13
Mean (ha)	0.83	0.50	0.45	0.63

*proportion by percentage in parenthesis.

The area under pigeonpea was 1.0, 0.6 and 0.3 ha in Machakos, Kitui and Embu, respectively. Maize occupied 20.7 percent of cropped area, pigeonpea 19.6 percent, beans 18.7 percent and cowpeas 14.5 percent. No other crop occupied more than 10 percent of area under crops.

Table 3 indicates that the major crops in the three districts are maize, cotton, pigeonpeas, beans, coffee, sunflower and fruits. Although cowpeas are considered drought tolerant they are grown on a smaller area and possibly by few farmers. Finger millet and sorghum are important in Kitui. In this district, the area under finger millet was 7.3 percent whereas 7.0 percent was under sorghum. In Machakos, 5.7 percent is under finger millet and 5.1 percent under sorghum. Due to frequent droughts, few farmers grow vegetables and fruit. Dolichos are popular in Embu.

PIGEONPEA FARMING

Mbatia and Kimani (1987) found that 86, 92 and 91 percent of farmers in Machakos, Embu and Kitui districts, respectively, have heard and grown improved pigeonpea. The majority of farmers interviewed stated that they grew improved pigeonpea every year. Only 7 percent of farmers interviewed did not grow improved pigeonpea yearly.

Time of planting

The breeding program of pigeonpeas in Kenya started in 1976/77. The objective was to improve existing traditional pigeonpeas that farmers have been growing for years. The time of planting is one of the key determinants of yield. Since the rainy season is short in semi-arid areas and moisture is often limiting during the season, timely planting is crucial for good yields. However, early planted short duration pigeonpeas often suffer severe insect damage. Late planting often gives poor yields. The recommended time of planting is late September and the whole month of October, i.e. short rain season. The highest percentage (70%) of farmers indicated that they planted their seeds at this time. However, some farmers also plant during the long rain season in March and April.

Types and sources of pigeonpea planted

The farmers indicated that they planted either Katumani or NPP 670 seeds and traditional pigeonpea. The NPP 670 is commonly known by farmers as Katumani because it matures early like the composite maize variety known as Katumani. The distribution of NPP 670 seeds planted by farmers was 75.8 percent in Machakos, 36.6 percent in Kitui and 72.6 percent in Embu.

It was the interest of this study to explore how the farmers came to know about improved pigeonpeas (NPP 670). Table 4 shows that 50, 45.6 and 59.3 percent of farmers in Machakos, Kitui and Embu, respectively got information on NPP 670 from agricultural extension officers.

Table 4: How farmers first heard about improved pigeonpeas.

Source	Machakos	Kitui	Embu	All
		%		
Neighbour	25.9	17.6	23.0	22.9
Extension officer	50.0	45.6	59.0	51.5
Radio	0.5	0.8	-	0.4
Neighbours field	9.9	7.2	5.9	8.1
Farmers training centre	7.1	1.6	-	3.6
Other	1.4	2.4	2.2	1.9
NA*	4.2	24.8	9.6	11.7

*NA - Not applicable; i.e. those farmers who stated that they have not heard about the improved pigeonpeas.

The source of the first improved pigeonpea seed which the farmers planted is shown in Table 5.

Table 5. Sources of the first improved seeds planted by farmers. (1983-85) [see also Tables 9 and 16]

Source	Machakos	Kitui	Embu	All
Local market	6.1	13.8	5.3	8.4
Local shop	9.1	13.8	8.5	10.5
Neighbour	22.0	17.2	47.9	29.0
Agricultural officer	47.0	55.2	26.6	42.9
Farmers training centre	2.3	-	-	0.8
University researchers	12.9	-	11.7	8.2

In all three districts, 43 percent of the farmers got their seeds from agricultural officers, 29 percent of the farmers acquired their seeds from neighbours and about 8 percent obtained seed directly from researchers at the University of Nairobi. In Embu the majority of the farmers (47.9 percent) got their seeds from their neighbours. The local markets and shops ranked fourth as a first source of improved pigeonpea seeds.

Farmers were asked to state the reasons why they did not grow the new varieties of improved pigeonpea. Table 6 shows that 14 percent of farm stated that the improved pigeonpeas were not good. They argued that improved pigeonpea varieties required a lot of labour for spraying against diseases and insects, were too short and not good for firewood. Only 4 percent of farmers stated that the seeds were too expensive, and less than 2.0 percent had no money to buy the seeds. About 8 percent of farmers stated that seeds were not available. Mbatia and Kimani (1987) found that 40.7, 42.1 and 35.7 percent of farmers in Machakos, Embu, and Kitui respectively stated that seeds were not available.

Table 6: Reasons given by farmers for not growing improved pigeonpea

Reasons	Machakos	Kitui	Embu	All
		%		
Seeds not available	9.4	9.6	4.4	8.1
Seeds too expensive	4.7	4.0	3.0	4.0
Cash not available	1.4	2.4	1.5	1.7
Varieties not good	5.2	28.0	8.1	14.4
Other reasons	0.5	4.8	8.1	6.3
NA*	68.4	50.4	74.8	65.5

*NA - not applicable i.e. farmers who did not have any problems in getting seeds.

SOURCES OF IMPROVED PIGEONPEA SEEDS

The farmers obtained seeds for planting from various sources such as own seeds, buying seeds from neighbours or from the markets. Table 7 shows where the farmers obtained seeds during the period 1987-89. The major source of improved pigeonpea seeds were farmers own harvested seeds or from the extension officers. The extension officers were supplied with seeds by the researchers from the pigeonpea project, University of Nairobi. There is a steady increase from year to year of farmers who planted their own seed.

Table 7: Sources of improved pigeonpea seeds in Machakos, Kitui and Embu districts: 1987 to 1989.

Source	1987	1988	1989
		percent	
Own seeds	20.3	33.1	37.3
Neighbours	7.6	8.1	6.6
Local market	5.7	3.8	1.7
Extension officers	16.7	12.3	12.5
Friends/relatives	0.8	0.4	-
Other	1.3	0.8	1.1
NA	47.3	41.5	40.9

The traditional pigeonpea seeds planted by the farmers were available at home, or purchased from markets and neighbouring farmers. Nearly all the farmers in the three districts where the survey was carried out indicated that they plant traditional pigeonpea every year. In 1987, 71.4 percent of the surveyed farmers planted seeds of their own traditional pigeonpeas. This increased to 74.8 percent in 1988 and to 78.2 percent in 1989. It is therefore, safe to say that for both improved and traditional pigeonpeas, the farmers used the seeds they harvested for planting. The buying of seeds from the market was insignificant in the three districts. The price charged for improved pigeonpea seeds for planting is Kshs. 17.00 per kg.

The farmers were asked whether they were able to buy as much improved seeds as they would have liked. Ninety one percent of the sampled farmers responded they were able to get as much seeds as they wanted. A very small number of the farmers, 2.3 percent, responded that the seeds were not available. Those who stated cash was a problem were 4.4 percent of the total surveyed. It should be concluded that the improved pigeonpea seeds were available and that the farmers were able to get seeds for planting.

PRODUCTION AND CONSUMPTION OF PIGEONPEA

In Machakos, Kitui and Embu districts, the majority of the farmers grow pigeonpeas either for home consumption or for sale in order to earn some income.

Table 8 : Production of improved pigeonpea seeds and number of farmers growing them, 1983 - 1989.

Year	Seed production* (t)	Number of farmers#
1983-84	3	5,000
1984-85	10	20,000
1985-86	11	30,000
1987-88	16	50,000
1988-89	20	over 55,000

* Based on seed packets sold to individual farmers by MIDP or pigeonpea project.
# Ministry of Agriculture estimates.

Table 8 shows the amount of seeds of improved pigeonpea cultivars produced and sold to farmers and estimated number of growers. The figures indicate that amount of seed produced increased over sixfold while number of farmers increased tenfold during the seven year period. It should be noted that total amount of seed produced is likely to be much higher since the seed produced and sold by individual farmers or private seed companies was not included. Much of the seed produced by the project or MIDP was distributed through the extension officers, co-operative unions or sold directly to farmers. Data on Table 6 however indicated that 28 percent of seeds planted by farmers in 1987 came either from their own seed, local markets or from friends or relatives. In 1989, this figure rose to 45.6 percent. This is further supported by data on Table 5 which indicated of total seeds first planted by farmers, 48.7 percent originated from local markets, local shops or neighbours.

According to the Ministry of Agriculture report (1986) the area under pigeonpeas in Eastern province rose from 70,277 hectares in 1985 to 93,238 hectares in 1986, an increase of 32.7 percent. Production for the same period rose from 37,608 to 54,070 tonnes, an increase of 44 percent. This increase was attributed to rapid adoption of cultivar NPP 670 and favourable market conditions (Annual report, Ministry of Agriculture, 1986).

Over 85 percent of farmers in Machakos, Kitui and Embu stated that they did not purchase any pigeonpea for home consumption. In 1989, 11.7 and 12.6 percent of the farmers in Machakos and Embu, respectively bought less than 10 kg of pigeonpea for home consumption. There were relatively few farmers in Kitui who bought pigeonpeas for home consumption compared to the other districts.

Cooking of pigeonpea

The amount of time food takes to cook depend on many variables such as firewood used and composition of grain being cooked. On the basis of their cooking experience, farmers were asked which pigeonpeas cooked faster, improved or traditional pigeonpeas. Table 9 shows the reaction of farmers to this question. Seventy percent of sampled farmers (mainly women) answered this question. Forty-five percent indicated that the improved variety cooked faster than traditional one. An earlier study by Mbatia and Kimani (1987) found that improved pigeonpea took shorter time to cook than traditional one.

Table 9: Responses of surveyed farmers as to which cooks faster, traditional or improved pigeonpea varieties in percentage

Relative cookability	Machakos	Kitui	Embu	All
No Difference	33.3	40.7	23.7	31.4
Improved variety	27.6	53.7	68.9	44.9
Traditional variety	39.1	3.6	76.4	23.7

According to 70 percent of surveyed farmers who had cooked and tasted both varieties, 40 percent indicated that traditional pigeonpeas tasted better than improved variety. Thirty-four percent of farmers stated that the improved variety tasted better than the traditional one. The taste preferences of the farmers are shown on Table 10.

Table 10: Taste preference of surveyed farmers in percentage

Preferred variety	Machakos	Kitui	Embu	All
Improved variety	17.1	25.9	40.6	26.0
Traditional variety	56.0	11.1	27.4	39.7
Both are about the same	26.9	63.0	32.0	34.3

PURCHASE OF IMPROVED PIGEONPEA SEED FOR PLANTING

The improved pigeonpea seeds for planting by the farmers could be acquired from various sources such as farmers using their own seeds or purchasing. It was found that 66.6 percent of farmers purchase seeds, and that the rest used some of their own seeds from previous crops and bought some.

Table 11 shows the place where the farmers said they would purchase the improved pigeonpea seeds. About 53 percent of the farmers would buy the improved pigeonpea seeds from agricultural officers and 16 percent indicated that they did not know where to buy the improved pigeonpea seeds.

Table 11: Places where farmers said they would buy improved pigeonpea seeds for planting (in percentage)

Place of purchase	Machakos	Kitui	Embu
Local shop	3.9	2.2	-
Local market	8.3	6.4	3.7
Agricultural office	48.5	28.0	58.8
From neighbour	9.8	2.2	10.0
Co-operative store	12.3	1.6	-
Do not know	10.2	67.0	-
NA	-	-	8.15

In 1988, 97 percent of farmers planted pigeonpeas. The varieties of pigeonpeas planted by farmers are shown in Table 12. The traditional pigeonpeas were planted by 37.9 percent of sampled farmers. Those farmers who planted only improved varieties amounted to 6.4 percent. Both varieties were planted by 55.7 percent of the farmers.

Table 12: Type of pigeonpea planted by farmers in 1988 in percentage

Type	Machakos	Kitui	Embu	All
Traditional pigeonpea	29.6	66.1	26.3	37.9
Improved pigeonpea	1.4	4.3	16.7	6.4
Both varieties	69.0	29.6	57.1	55.7

Most of the farmers surveyed grew pure stands of the improved pigeonpeas but the traditional pigeonpeas were usually intercropped. The improved pigeonpeas were hardly intercropped with traditional pigeonpeas.

HARVESTING AND STORAGE OF PIGEONPEAS

Quantity of pigeonpea harvested

Table 13 shows the quantities of pigeonpea harvested in 1988 season. In all the areas, more than two thirds of the farmers harvested less than or up to 5 bags. On average farmers in Machakos harvested 4.7 bags, in Kitui 4.3 bags and 2.3 bags in Embu. Machakos is the leading producer of pigeonpeas, followed by Kitui and Embu. A bag of pigeonpeas weighs 120 to 130 kgs.

Table 13: Quantities of pigeonpeas harvested by farmers (in percentage) in 1988 in three districts

Bags	Machakos	Kitui	Embu	All
Upto 5	73.1	75.7	88.9	78.1
5.1 - 10	20.9	17.1	8.5	16.6
10.1 - 15	3.0	4.5	1.7	3.0
15.1 - 20	2.0	1.8	0.9	1.6
20.1 - 30	0.5	0.9	-	0.5
Over 30	0.5	-	-	0.2

Storage of Pigeonpeas

The method used in storing grain is very vital in reducing damages caused by rotting, insects and rodents. Table 14 shows the methods used by farmers. Ninety-five percent of farmers used bags in storing pigeonpeas. This is the most common method employed for storing produce by farmers.

Table 14: Methods used by farmers to store pigeonpeas after harvest (in percentage)

Storage method	Machakos	Kitui	Embu	All
In bags in store	97.2	88.3	96.0	94.5
In open in a store	2.8	11.7	0.8	4.6
Outside (house or store) But covered	-	-	0.8	2
Others	-	-	2.4	7

The surveyed farmers reported that they treated harvested pigeonpeas during or before storage. Ninety-six, 86 and 82 percent of farmers in Machakos, Kitui and Embu districts, respectively treated their produce before or during storage.

Forty-six percent of the surveyed farmers stated that their produce was damaged during storage in all three districts. The rest reported that their produce was not damaged. In all districts 87.6 percent of farmers stated that the major damage was due to insects. Forty-eight percent stated that the damage to the produce was very little. This indicates that about half of the farmers employed proper methods of storing produce.

MARKETING OF PIGEONPEAS

The primary objective of growing pigeonpeas or other crops by farmers could be either for home consumption or to sell in order to earn some income. The income is used for buying essential household goods such as food, and clothing and for services such as school fees, medical services, farm inputs and others. The surveyed farmers were asked if they had sold pigeonpea harvested in the 1988 season or they intended to sell in future. The farmers who responded positively to the question were 56.4 percent. About 54 percent of the farmers indicated that they had no intention of selling pigeonpeas in future.

The time for selling pigeonpea is very important. At harvesting time there is more pigeonpea in the market than there is demand. This causes the price to be low. In 1988 season, the price was Kshs 3.00 per kg. Table 15 shows the various times when the farmers put pigeonpea for sale. Thirty-four percent of farmers sold the pigeonpeas at the time of paying school fees. The demand for money is very high during this time since this is the only source of income for many farmers. At harvesting time 36.8 percent of farmers sold pigeonpeas. The majority of the farmers (86.2 percent) who had no intention of selling pigeonpeas stated that they only had enough to meet their domestic consumption. This confirmed an earlier observation that farmers do not buy pigeonpeas for their domestic consumption.

Table 15: Times when farmers normally sell pigeonpeas (in percentage)

Time	Machakos	Kitui	Embu	All
Harvesting time	26.4	65.6	30.2	36.8
Planting time	13.1	33.0	1.6	7.9
Paying school fees	35.7	19.7	42.9	33.6
Buying food	8.5	3.3	14.3	8.7
Others	16.3	8.1	11.0	13.0

Market outlets for pigeonpeas

Table 16 shows the various marketing outlets which farmers used to sell pigeonpeas. In Machakos and Kitui the major buyers were traders. The traders normally operate shops in nearby shopping centre where the farmers take their produce. In Embu 56.6 percent of farmers sold pigeonpeas at the markets. The buyers include many consumers who buy small amounts and also traders who buy and transport the produce for selling in urban centres such as Nairobi and Mombasa. The market days are normally twice per week.

Table 16: Marketing outlets for pigeonpeas in percentage of farmer response

Outlet	Machakos	Kitui	Embu	All
Neighbours	8.9	1.2	21.7	10.6
Nearby shop (traders)	67.9	74.7	10.8	51.1
Local markets	16.8	19.3	56.6	30.9
People who come to buy (Merchants)	4.7	1.2	10.8	5.6
Pigeonpea project	1.6	3.6	-	2.6
Others	-	-	-	

Table 16 also shows that only a small fraction of the pigeonpea grain is sold back to the project as seed.

Prices for pigeonpeas

The price for pigeonpea is not controlled by the government. Normally by word of mouth, farmers are aware of going price at farm gate, at the shopping centre and also at the market. The prices are low during harvesting time and high during planting time. The price of pigeonpeas, either improved or traditional, for home consumption is between Kshs 3.00 to Kshs 4.00 per kg. The price paid for improved pigeonpea seeds to farmers contracted by the pigeonpea project is Kshs 8.00 per kg.

The farmers were asked to react to the price they received i.e. whether it was good, poor or reasonable. The farmers' responses are shown in Table 17. About 23 percent of the farmers felt the price was good, and 31.8 percent of the farmers considered the price to be poor.

Farmers stated that they experienced problems such as transportation in getting their produce to the market. They noted that transportation was unavailable and when available it was too expensive. Over 50 percent of farmers in all three districts indicated that they had problems getting their pigeonpeas to the market.

Table 17: Reaction of farmers to prices paid for pigeonpea seeds (in percentage)

	Machakos	Kitui	Embu	All
Very good	4.4	8.7	1.1	4.4
Good	22.2	7.2	36.7	22.9
Fair	20.2	7.2	25.6	19.1
Poor	27.6	47.8	28.9	31.8
Very poor	25.6	8.7	7.8	21.8

EXTENSION

The farmers received information on new technology through radio broadcasts, extension workers, field days, through other farmers and during agricultural shows. They stated that they were aware of radio programs on farming. Although these radio programs on farming are in Swahili language, some of them are also broadcast in local languages. The extension workers communicate to the farmers in local languages during farm visits and on field days. Apparently there were no problems of communicating new technology to farmers. This has contributed to rapid adoption of the new early maturing pigeonpea.

DISCUSSION

Seed multiplication mechanisms

The project has pursued various mechanisms for seed multiplication with varying degrees of success. Multiplication of seed on leased land at Kibwezi was initiated by the project in 1983. All the operations were carried out by the project personnel and good quality seeds were obtained due to close supervision, suitability of land and availability of irrigation. However, the land rent became exorbitant which implied higher seed costs. This was considered undesirable since the aim of the project was to provide growers with good quality seed at a price they could afford. Besides seed production operations made heavy demand on limited project personnel. The Kibwezi area, Machakos district, is thinly settled and much of it being bush, wild animals often damaged the seed crop.

Multiplication of seed by MIDP worked well except for the frequent changes in personnel. Being an externally funded project, it had a limited life span and hence not a sustainable method of seed multiplication.

The small-scale farmer contracts worked best. Farmers were ready to produce as much seed as required so long as there was a good market for their produce and good prices. This, however, requires a revolving fund for purchase of chemicals, provision of foundation seed, gunny bags and on-the-spot payment. It also requires close supervision and functioning arrangements for marketing of the seed. Loan of chemicals and foundation seed were easily recovered at the time of seed purchase preferably in the form of produce. For the farmers it was a crucial income generating enterprise. It has the greatest potential.

The district based pilot schemes operated along the same lines as farmers contracted by the project. This scheme proved workable as long as funds were available for the purchase of inputs or to repurchase the seed from farmers. The major drawback was that funds from seed sales were paid to the treasury and not ploughed back into the scheme, other than through the normal allocations for each district.

Although it was generally felt that multiplication of seeds by women's groups had great potential, this potential was not fully realised. The project provided about 240 kg of foundation seed in 1989, having jointly worked out a plan of operation with the women's bureau (Ministry of Culture and Social Services) which represented women's groups. However, the seeds were distributed to individual members. Although it is too early to make firm conclusions on this mechanism, proper management and co-ordination is essential.

As indicated earlier, private companies are principal multipliers and distributors of seed in Kenya. The results of this survey clearly supported their view that demand for seed is not firm. Data on Table 7 showed that the number of farmers using their own improved seed increased progressively from 1987 to 1989. This was also true of seeds of traditional pigeonpea varieties. The majority of the farmers do not buy pigeonpeas for their own consumption. This pattern is likely to persist so long as open or self-pollinated cultivars can be grown year after year without any serious deterioration in yield. Yet there is demand for seed of new varieties and especially after drought years. The survey results indicated that 66 percent of the farmers purchased seed of improved cultivars for planting. There is urgent need to quantify this demand and provide more market information.

It is in view of these constraints that the government-owned Kenya Agricultural Research Institute has agreed to contract private seed companies to produce seeds of semi-arid lands, pigeonpea included.

Dissemination mechanisms

The results indicated the largest proportion of the seeds of the improved varieties was disseminated initially through extension agricultural officers and farmer-to-farmer (Table 5). Most farmers also learnt about these cultivars through extension officers and their neighbours (Table 4). In subsequent years, the farmers used their own seeds for planting (Table 7). Direct purchases from the University accounted for only 8.2 percent of the seed distributed. This indicates clearly that agricultural extension officers and farmer-to-farmer sales were the most crucial mechanisms of disseminating the seed of the new cultivars. It may be inferred that farmers regard extension officers as their primary source of information on new technologies. This is further supported by data presented on Table 11. Asked where they would buy seeds of improved cultivars, 59 percent of the farmers said from extension officers and 10 percent from their neighbours.

The large number of responses indicating agricultural offices as sources of information or new technology could be attributed to the fact that there is at least one extension officer at grassroots or locational level. These officers are responsible for visiting farmers in their homes regularly or inviting them to field days or barazas (meetings) where farmers are informed of the latest information relating to agriculture, and which also provide a forum where farmers can air their views. The project involved these officers in most of its field operations. This outlet should be exploited in future seed dissemination mechanisms.

Local shops and markets and co-operative stores can also be used effectively, for seed distribution, since they are to be found in the remotest of the places. It is the combined effect rather than any one single channel that contributed to the rapid dissemination of seeds of the new pigeonpea cultivars. These mechanisms should be utilised fully regardless of the institution multiplying and distributing the seed. Co-operative stores such as KGGCU which have branches in all major urban areas and distributes other farm inputs have a great potential in seed dissemination.

The demand for pigeonpea seeds has been high. Mbatia and Kimani (1987) showed that farmers had some difficulties in getting seeds. They reported that 40.7, 42.1, and 35.7 percent of farmers in Machakos, Embu, and Kitui, respectively stated that seeds were not available. However, in the present study, only 8 percent of the farmers stated that seeds were not available. Ninety-one percent of the sampled farmers responded that they were able to get as much seeds as they wanted. This confirmed that seed dissemination mechanisms were effective and farmers had access to seeds of the new pigeonpea cultivars. The project has received some requests for seeds of new varieties from Malawi, Tanzania, Uganda, Zimbabwe, Somalia, Pakistan, India and Sudan. Most of these requests have been met.

Adoption of improved pigeonpeas

The improved pigeonpea varieties have been widely adopted by farmers in Machakos, Kitui and semi-arid areas of Embu district. Mbatia and Kimani (1987) found that 86, 92 and 91 percent of farmers in Machakos, Embu and Kitui districts, respectively, have heard and/or grown improved pigeonpeas. In the present study, over 88 percent of sampled farmers have heard about improved pigeonpea mainly through extension officers and neighbouring farmers. Sixty-four percent of surveyed farmers have grown the improved cultivar NPP670 and about 54 percent of them grow it yearly. The rapid adoption of the new cultivars is, in part, attributable to their desirable traits, availability of seeds, reasonable prices and rapid dissemination of information by extension officers. Information on the new technology was relayed to farmers through agricultural extension officer, FTC's and from farmer-to-farmer. Radio played a relatively small role in disseminating information on the improved cultivars.

Asked what they liked about the improved cultivars, 52 percent of the farmers cited its early maturity, 20.1 percent high yields when sprayed, 9.1 percent stated that it can be harvested twice per year, 10.8 percent, better taste, and 3.4 percent better market prices. Other reasons that were cited include faster cooking (1.5 percent), short stature hence easy to spray (1.4 percent), and big seeds (1.1 percent). Among the dislikes were : insect susceptibility (30.4 percent), need to spray heavily (33.1 percent), diseases (15.1 percent), heavy labour demand (8.1 percent), expensive seeds (4.7 percent) and low yields (5.1 percent).

Seed marketing

In 1988/89 56 percent of the farmers indicated that they intended to sell part of their pigeonpea produce. Over 50 percent of the farmers indicated that they had problems getting their pigeonpeas to the market. The most serious problem was transportation. Either the transport was not available or if available was too expensive. Others cited long distances to the markets. Perhaps the most potential solution to this problem is to organise a growers association so that the produce can be collected at several points and delivered to markets. Such associations exist for vegetable, coffee and tea growers. They assist the farmers in locating demand areas and negotiating better prices for their produce. Little market information is available on pigeonpea domestic or export markets. The future of increased pigeonpea production in Kenya lies in quest for market information, organised marketing and its linkage to production.

EXPERIENCES IN DRYLAND SEED PRODUCTION AND DISTRIBUTION

Although the pigeonpea project has made efforts to supply seed to farmers in collaboration with MIDP, several problems have been experienced. Seed production is an expensive and time consuming activity. Seed fields require full time staff to manage the fields and carry out all operations, ranging from land preparation to harvesting, cleaning, dressing, packaging, storage and distribution of seed to the farmers. Facilities for these operations are necessary since hired equipment from private companies means that seeds have to be sold at high prices to break even. This is complicated by the need to guard the crop from wild animals and theft, land leases and use of costly chemicals to ensure seed of high quality is delivered to farmers. Resource poor farmers in semi-arid lands cannot afford costly seed and this forces some to use unimproved seeds. Fortunately, unlike hybrid seeds, pigeonpea seed derived from open pollinated cultivars can be replanted for a few years without serious decline in yields. Rough terrain and impassable roads make it difficult to deliver seeds closer to farmers in the more remote areas. Seeds have to be sold at subsidized prices to ensure that new varieties are adopted by as many farmers as possible.

Dryland Seed Production and Distribution Committee

This committee was formed to assist in developing strategies for multiplication and distribution of seeds of dry land areas. It arose out of the realization that commercial companies were hesitant to multiply and distribute seeds for semi-arid lands except for maize, beans and sorghum which is carried out by Kenya Seed Company. The committee met at the National Dryland Farming Research Centre, Katumani on November 11, 1984. The purpose of the meeting was to work out a strategy for seed production and distribution for the semi-arid areas. Among the institutions represented were government agricultural research stations, district agricultural officers of Embu, Kirinyaga, Kitui, Machakos, Baringo, Kenya Freedom from Hunger Campaign, National Seed Quality Control Service (NSQCS), Kenya Seed Company, East African Seed Company, MIDP and the University of Nairobi. The main points made during that meeting were:

1. There is a critical need to provide good quality seeds to farmers and to ensure a supply of seed after bad cropping season.

2. Although official regulations require that varieties have to pass the National variety performance trials for three years before official release, under the present circumstances, good material from breeders should be multiplied to provide the farmers with seed, to create awareness and get a feedback to perfect the research work.

3. Although the Kenya Seed Company was willing to multiply seeds for dryland crops, the exercise must be profitable. The company's experience with multiplying GLP beans was unpleasant due to uncertain demand. The majority of the farmers bought certified seed only after bad seasons. The East African Seed Company would multiply the seed only if they were sure the demand would be high enough.

It was suggested that :

i) seed prices should be subsidized to ensure farmers buy every season and the company could produce large quantities regularly.

ii) The National Cereals and Produce Board give premium prices for pure varieties. This would encourage farmers to buy more seed to produce pure varieties. It was concluded that this was a policy matter that needed further discussion at the ministerial level.

4. It was suggested that farmers should be given small quantities of seed to start them off and educate them that most of the dryland crop seed can be grown for more seasons and that after 2-3 seasons they will have to buy new seed. However, in bad seasons, farmers still need seeds to buy.

5. Commercial seed companies were not interested in vegetatively propagated crops such as potatoes and cassava. Multiplication of these should be left to breeders, extension services, farmers or institutions.

6. No institution appeared to have resources to multiply and organize seed distribution for arid lands. Apparently the only viable solution then was for each district to multiply and distribute seed to their farmers. The seeds unit at the National Dryland Farming Research Centre, Katumani had run out of resources to multiply seed and any new cultivars were given to MIDP.

7. General rules for seed multiplication for cowpeas, green grams and pigeonpea were discussed. The commercial growers as well as DAO offices/projects were free to apply to grow the seeds, but applicants should have seed multiplication facilities.

Since demand for seed in marginal areas fluctuates seasonally with amount and distribution of rainfall, organized seed production and distribution was difficult to carry out. This activity would have to be done at the institutional level. The project researchers proposed small seed production pilot projects based at the district level. In this scheme small quantities of seed is sold to farmers who are encouraged to reserve some of the harvested seed for next planting and sale to neighbours. A few farmers are also contracted to produce seed to be purchased using a revolving fund. This seed is offered for sale through district agricultural offices and local shops to those farmers unable to produce enough seed in a previous season. The pigeonpea project will ensure that emergency seed stocks are available for each of the cultivars. In the long run, when a sufficient demand has been created commercial companies may be attracted. Seed production rules and regulations would be followed. It was generally felt that the Ministry headquarters should get more involved in seed multiplication and distribution for crops of semi-arid lands. Breeders should spend more of their time in cultivar development and research and less in seed production activities.

POLICY IMPLICATIONS AND RECOMMENDATIONS

The growing of improved and traditional pigeonpea is being carried out by small-scale resource poor farmers in semi-arid areas. This study has shown that the surveyed farmers have accepted growing improved pigeonpeas which are early maturing and of higher yielding capacity compared to that of traditional varieties. The farmers have experienced prevalent problems of insects and diseases as well as marketing. To help farmers overcome some of the constraints to increased production the following recommendations are suggested :

1. Research on breeding should continue - the farmers need cultivars which are resistant to diseases and insects. This requires more money, and, above all, team work of researchers with different scientific backgrounds such as entomologists, pathologists, among others (the project already has entomology, pathology, agronomy and breeding graduate students doing some work on these aspects).

2. The Ministry of Agriculture should take up the work of seed production and distribution. The pigeonpea project should concentrate on breeding, agronomy, socio-economics, entomology and pathology of pigeonpeas only.

3. Agents could be appointed by the government through the Ministry of Agriculture to carry out seed production and distribution. Some cost sharing mechanism should be worked out between farmers, agents and government so that seeds are produced at reasonable prices.

4. Extension workers should be educated regarding problems relating to improved pigeonpea and how farmers could select good seeds for planting.

5. In absence of seed agents good farmers in the area should be trained to produce seeds for planting. These farmers could be contracted to produce quality seeds.

6. A distribution system should be instituted to ensure that improved seeds reach farmers. Seeds could be distributed through agricultural offices, local shops, co-operative stores such as the Kenya Grain Grower Co-operative Union (KGGCU) chain stores, private seed companies and markets.

7. Agro-chemicals especially insecticides and spraying pumps should be made available to farmers at reasonable cost. The current prices for these are too high for most farmers. This issue needs urgent attention.

8. The marketing of pigeonpeas should be improved. Efforts should be made to organise farmers into groups so that costs of transportation and marketing their produce can be reduced. Better markets should be sought.

9. The Ministry of Agriculture should promote improved pigeonpea as one of the most drought resistant crop in the semi-arid areas.

10. The germplasm of improved pigeonpea varieties should be kept in national seed bank and be registered.

These policy recommendations are aimed at improving and increasing production of improved pigeonpea in semi-arid areas as the pigeonpea will continue to be a major crop in semi-arid areas. Therefore a joint effort between farmers, scientists, extension workers, politicians and policy makers is required to ensure sustainable production in agriculture. Good quality seeds must be provided to farmers to ensure sustainability.

CONCLUSIONS

1. The adoption of new varieties of pigeonpea has taken off very well in Machakos, Kitui and Embu districts. Over 88 percent of sampled farmers have heard about improved pigeonpea mainly through extension officers and neighbouring farmers. Sixty-four percent of surveyed farmers have grown the improved variety of pigeonpea i.e. NPP 670. Approximately 54 percent of sampled farmers grow improved pigeonpea yearly.

   The farmers got the first seeds for planting from the agricultural officers (41 percent) and 31 percent of farmers got seeds from neighbouring farmers. In Kitui district much of the seed sold to farmers by the agricultural officers originated from the pilot seed multiplication project with small amounts supplied direct by the project headquarters at the Department of Crop Science, University of Nairobi. In Machakos district seeds were jointly multiplied and distributed by Machakos Integrated Development Project (MIDP) and the pigeonpea project. In Embu district the seeds were supplied to agricultural officers by the project from project nurseries and contract farmers. In 1988, 33.1 percent of surveyed farmers planted their own seeds, 12.3 percent bought seeds from agricultural offices, and 8.1 percent of farmers purchased seeds from neighbouring farmers.

2. The improved pigeonpea seeds were available to the farmers. Of the farmers who were growing traditional pigeonpea, 78.2 percent planted their own seeds. The rest of the farmers purchased traditional pigeonpea seeds at the market.

3. The prevalent problems experienced by farmers growing pigeonpea were insects and diseases. The farmers also indicated that the price paid on pigeonpea was poor. The farmers had problem in taking the produce to the market. They stated that transportation was expensive and unavailable.

   The improved pigeonpea seeds are sold at Kshs 17.00 per kg. Seeds are packed in half kg packages. It is most likely that the improved seed bought from neighbouring farmers cost less than Kshs 16.00 per kg. In Karaba market, in Embu, improved pigeonpeas were selling at Kshs 12.50 per kg in 1988. Other donor agencies working in the semi-arid areas sold pigeonpea seeds at Kshs 8.00 per kilogramme.

4. Forty-five percent of sampled farmers reported that improved pigeonpeas cooked faster than traditional pigeonpea. About 40 percent of surveyed farmers responded that traditional pigeonpeas tasted better than improved pigeonpea. Other advantages cited by farmers about improved pigeonpeas include early maturity and that they harvest two crops per year. The traditional pigeonpeas are admired by farmers because of lower insect infestation and disease problems and also are a good source of firewood and fencing material.

5. The common method for harvesting pigeonpeas was cutting of entire crop and threshing by beating with a stick. This method was used by 61 percent of the surveyed farmers. Pigeonpea was stored in bags by 95 percent of farmers. The farmers also treated stored produce with agrochemicals.

6. The improved seeds for planting were supplied to agricultural offices by the Pigeonpea Project of the University of Nairobi funded by the International Development Research Centre (IDRC) Ottawa, Canada. The seed multiplication and distribution was carried out mainly by contracted small-scale farmers and the Pigeonpea Project. The seeds from contract farmers are purchased by the project, for sorting, dressing and packaging. Seeds are distributed through agricultural offices, local markets and shops and directly from project offices. Contract farmers also sell their seeds to their neighbours and in local markets. The amount of improved seed that has been produced and sold is difficult to quantify. A lot of seed is sold among farmers and from farmers to traders that is not recorded. The amount of seed produced by the project, MIDP and the pilot seed scheme in Kitui district increased from 3 tonnes in 1983/84 to over 20 tonnes in 1988/89. In the same period the number of collaborating farmers who received and planted this seed increased from 5,000 in 1983/84 to over 50,000 farmers in 1988/89. From this survey and 1987 socio-economic survey, it was estimated that over 64 percent of the farmers had grown the improved pigeonpea varieties. The population of three districts is estimated to be over 2 million people and an average household size of 5.3 persons i.e. 377,358 households (Jaetzold and Schmidt, 1979). It can be estimated that about 241,509 households (or farmers) have grown the improved pigeonpea cultivars (64 percent) which is about five times the recorded estimate.

   The area under pigeonpeas in Eastern Province (Machakos, Embu, Kitui, Marsabit and Meru districts) increased from 70,277 hectares in 1985 to 93,238 hectares in 1986 (an increase of 32.7 percent) according to the Ministry of Agriculture report (1986). Production in the province which accounts for 90 percent of Kenya's pigeonpea crop, rose from 37,608 tonnes in 1985 to 54,070 tonnes in 1986, an increase of 44 percent. The increase in hectarage and production was attributed to a good market and improved pigeonpea varieties.

7. The multiplication and distribution of pigeonpeas requires a lot of investment both human and capital. It is suggested that a division of labour is required, where the Pigeonpea Project of the University of Nairobi should concentrate on breeding and agronomic research while the Ministry of Agriculture should concentrate on seed multiplication and distribution. For instance the Ministry of Agriculture could contract private seed companies to multiply the seed. The seed would then be distributed through the Agricultural Extension Officer, Co-operative union stores and shops among other channels. This appears to be a logical and efficient way of supplying farmers with the required improved pigeonpea varieties and other crops of semi-arid areas.

REFERENCES

1. Akinola, J.O. and P.C. Whiteman. 1972. A numerical classification of Cajanus cajan L. Millsp accessions based on morphological and agronomic attributes, Australian J. Agric. Res. 23: 995-1005.

2. Kimani, P.M. 1985. Pigeonpea: a multipurpose woody legume for Agroforestry. Paper presented at the 5th USAID-ICRAF Workshop for Agroforestry Scientists 4-23 November, 1986. Nairobi, Kenya.

3. Kimani, P.M. 1987. Pigeonpea breeding in Kenya. Objectives and Methods: In ICRISAT: Pigeonpea Scientists' Meet. Proceedings: 2-5 June 1987 pages 5-28. ICRISAT, Hyderabad, India.

4. Kimani, P.M. 1988. Pigeonpea Improvement in Kenya. Final report in Phase III. International Development Research Centre, Ottawa, Canada. 1988. 231 p.

5. Kimani, P.M. 1989. Pigeonpea Improvement in Kenya. Second report of Phase IV. IDRC, Ottawa, Canada. 111 p.

6. Mbatia, O.L.E. and P.M. Kimani. 1987. Socio-economic survey of pigeonpea farmers in Machakos, Embu and Kitui districts in Kenya. IDRC, Ottawa, Canada/University of Nairobi, Pigeonpea Project, Nairobi, Kenya.

7. Ministry of Agriculture. Annual reports, Eastern Province (1982-1989).

8. Okiror, M.A. 1986. Breeding pigeonpea (Cajanus cajan L. Millsp.) for resistance to Fusarium wilt (Fusarium udum L.). Ph.D. thesis. Department of Crop Science, University of Nairobi, Kenya.

9. Onim, J.F.M. 1981. Effects of two population improvement methods on grain yield of pigeonpea composite populations in Kenya. Euphytica 30: 271-275.

10. Onim, J.F.M. 1983. On farm testing of improved pigeonpea (Cajanus cajan L. Millsp) cultivars in Kenya. In Kirkby, R.A. (ed). Crop Improvement in Eastern and Southern Africa: Research Objectives. IDRC, Ottawa, Canada.

SEED PRODUCTION ADOPTION AND PERCEPTION AMONG KAREN AND HMONG FARMERS OF NORTHERN THAILAND

Uraivan Tan-Kim-Yong
Manee Nikornpun

Faculty of Social Sciences and Faculty of Agriculture
Chiang Mai University, Chiang Mai, Thailand

ABSTRACT

The International Development Research Centre, Canada has funded Vegetable Seed Production project at Chiang Mai University, Thailand over the past eight years. Main objectives have been to establish vegetable seed production in northern Thailand to replace opium production and at the same time develop improved varieties of some vegetables for farmers.

Target groups of farmers for seed production are mainly hill tribe farmers. Vegetable seed production of three crops, Chinese radish, leaf mustard, and Chinese cabbage were introduced to them in 1986. Economic data showed that production of Chinese radish seeds was the most beneficial to farmers among the crops tested. The crop was introduced to two ethnic groups: Karen and Hmong on Inthanon mountain in Chiang Mai.

Generally, the Karen grow mainly paddy rice, some cabbage, and few home garden native crops, while the Hmong do not grow rice but produce cabbage all year round, fruit trees, strawberries, flowers and some new crops. Both groups adopted Chinese radish seed production. The Hmong group is more interested than the Karen in such production. The Karen produced the seeds on paddy land, transplanting seedlings and furrow irrigation, while the Hmong produced the crop on sloping land, direct seeding, and sprinkler irrigation. The methods of the Hmong were successful, while the Karen failed. The failure was probably a result of the long distance between home and the work field, home and extension stations, lack of attention from extensionist, ethnic background, lack of labour, and lack of incentives for adoption. The Hmong have been doing well in production and the number of interested farmers has been increasing in recent years.

INTRODUCTION

Vegetable seed production and varietal development is a process which includes experimenting, testing, and introducing an improved variety to local communities or farmers. The International Development Research Centre, Canada has supported these activities for the past eight years. The five crops undergoing varietal improvement and seed production are: Chinese radish, Chinese cabbage, leaf mustard, lettuce and sweet corn. Among these crops, the most advanced material ready for farmers are the lettuce and sweet corn varieties. Improved varieties of lettuce and leaf mustard are open pollinated varieties. Improved varieties of Chinese radish, Chinese cabbage, and sweet corn are F1 hybrid varieties. The improved varieties of three crops: lettuce, leaf mustard, and sweet corn in the project have been tested in many locations in Chiang Mai for several years and they showed their superiority over locally grown varieties in yield, horticultural characteristics, and disease resistance.

Results from many trials and locations showed that some of our improved lettuce varieties were better than some locally grown varieties in yield, heading percentage, head characteristics, and tipburn tolerance. However, extension of seed production of this crop has not yet started. We were not able to get a market for head lettuce seeds. The use of seeds in the country is very small and there is no big commercial scale distributor. Most of the seeds used are imported by two projects: the Royal King's project and Kao Kor project. Besides market problems, handling of seed production of head lettuce is rather more difficult than other vegetables in our project and the price of seeds (imported price) is not very attractive to farmers.

Varieties of leaf mustard were tested in our experiment stations for more than ten trials in winter and rainy seasons. They showed superiority over locally grown leaf mustard in yield, heading percentage, head characteristics, and low bolting percentage. However, these improved varieties have not yet been released because they will be taken another step to be F1 hybrid varieties.

F1 hybrid sweet corn varieties have been improved in the project. Two outstanding varieties were tested in comparison with commercial open pollinated and F1 hybrid varieties. They showed superiority in qualities and uniformity of the products over the commercial varieties in our varietal trials. They were tested by a few seed companies and some government institutions. Results showed that the varieties are superior in yield, uniformity, and horticultural characteristics of cobs and kernels. They were further tested for babycorn products in our trials and by other companies. The varieties are very outstanding in quality of babycorn and uniformity. Therefore, they can be used for sweet corn as well as babycorn production.

This study will emphasize seed production mechanisms only. We have successfully developed commercial Chinese radish seed production in several sites of northern Thailand. Originally, three crops, Chinese cabbage, leaf mustard, and Chinese radish, were introduced but only one crop has been adopted. Previous studies indicated that seed production could be successful on both slope land and irrigated fields at the altitude above 1,000 metres. A transfer of the seed production technology has been carried out by the project since 1987 after two years of testing by few farmers. The project scientists and extensionists worked with farmers to produce the seeds. The team also observed and documented the results of farmers' fields and management. The project aimed to test on different land types and management methods — one was on paddy land, transplanting seedlings, and using furrow irrigation; the other was on slope land, direct seeding, and using sprinkler irrigation.

A transfer of new technologies has often involved the project in technological aspects, and recently we have paid attention to sociocultural aspects. It has been recognized that new technologies cannot be introduced into local communities unless people are willing to change. The scientists of the seed production project found that farmers tended to accept only Chinese radish, and the cropping was more successful under the conditions of slope land and with management of Hmong ethnic farmers. The failure of seed production has been observed under the conditions of paddy land and with management of Karen ethnic farmers. Clearly, this specific scheme is socially complex, especially when adopters or seed growers are ethnic communities with variable cultural practices. Differential patterns of seed production between Hmong and Karen farmers illustrate many factors that influence farmer's acceptance. In understanding the contributing factors it is rather unrealistic to expect that the answers will be provided through any single social sciences study. The present stage of seed production research of

CMU requires the scientists to be aware of reasons for acceptance, and to establish ways to inform and teach farmers. A delivery system of inputs to farmers is also important, and needs to be considered.

Key questions raised in several studies on adoption are: who is most likely to adopt new agricultural technology, what is likely to bring about variation in the extent and timing of the adoption, what are appropriate mechanisms for working with farmers, and what are the problems confronting the adopters?

The principal purpose of this study is an exploratory investigation of the variation in adoption of seed production and perceived problems of Karen and Hmong adopters. We do not try to identify the causes of the particular adoption pattern. We believe that the two different adoption patterns mentioned cannot be adequately explained without reference to a wide array of factors in social and physical environments of two communities.

The field research was carried out during the wet season (May to September 1989) at Inthanon in the hill area of Chiang Mai province. Sixteen key informants or adopters were selected from both ethnic groups of Karen and Hmong. Both of them lived and cultivated in the villages at the altitude between 1000 metres to 1,500 metres. The researcher applied the methods of structured interview and observation for collecting sociocultural and economic data from the project villages to understand adoption and local situations, such as, production pattern, land use, labour use and exchange, agricultural knowledge and experience, acceptance of seed production, extension activities, production results, the problems perceived by seed growers, cooperation and group's activities, and interest and needs to produce seeds in the future, etc. In addition, the interviews were made with non-adopters, extensionists, and the scientists to obtain information about previous work on seed production research and constraints at initial stages.

COMPARISON OF KAREN AND HMONG VILLAGES AT INTHANON

The Karen is an ethnic minority well known as hill paddy farmers. These people are also engaged in several economic activities including a swidden agriculture, corn and buffalo raising, cash cropping, and wages employment. Being neighbours of Karen, the Hmong is considered a new migrant group which traditionally grows dry rice, corn, and opium in highlands of several provinces in the northern region. More than 85,000 Karen population and almost 15,000 Hmong population are now residing in hill villages of Chiang Mai.

Both Karen villages and Hmong villages are at their transition as more and more national and regional economic and sociopolitical intervention has increased during the last decade. Major development programs, such as crop substitution and highland agriculture development, have successfully drawn most of Hmong farmers into vegetable and fruit cropping systems, and the formal commercial sector of agriculture. At the periphery of development benefit, and of cash economy, the Karen continue to practice subsistence rice farming, and other traditional agriculture practices. Only small number of Karen farmers have tried new cropping of marketable varieties (Figure 1).

Figure 1. Crop Calendar of Karen and Hmong at Inthanon Mountain

CROP	WET SEASON							DRY SEASON				
	M	J	J	A	S	O	N	D	J	F	M	A
RICE (Karen)	x	x	x	x	x	x	x					
CHINESE RADISH SEED (Karen)							x	x	x	x	x	x
CHINESE RADISH SEED (Hmong)					x	x	x	x	x	x	x	x
CABBAGE (Karen, Hmong)	x	x	x	x	x							
CABBAGE (Hmong)	x	x	x	x	x	x	x	x	x	x	x	x
CORN FLOWER (Hmong)	x	x	x	x	x	x	x	x	x	x	x	x
STRAWBERRY (Hmong)						x	x	x	x	x	x	x
POTATO (Hmong)	x	x	x	x	x							
FRUIT TREE (Hmong)	x	x	x	x	x	x	x	x	x	x	x	x

More than fifteen villages are scattered within the Inthanon area. The seed production project selected two villages, Khun Klang (Hmong) and Pa Mon (Karen), for extension activity during 1987 to 1989. Karen farmers who adopted the seed production package were villagers from Pa Mon, Mae Klang Luang, and Sob Had. All Hmong adopters came from Khun Klang village where the demonstration plots were located. It was quite evident that Hmong farmers had some advantages over Karen who lived in villages some 10 to 20 kilometres distant.

From our economical studies on seed production in 1987, we were able to show farmers that seed production was beneficial to them (Table 1). Chinese radish, Chinese cabbage and leaf mustard were grown for seed production on slope and paddy lands. Seed yield and net income from Chinese radish seed production on slope land were the most satisfactory among the crops tested. Production of Chinese radish seeds either on slope land or on paddy land gave some profit to farmers. Then percentages of input in the cost of production of Chinese radish were analysed. Results are shown in Table 2. Major input of the cost was labour. Home labour and neighbour labour are seen as important factors in production.

Table 1. Seed yield, income, net profit of Chinese radish (CR), Chinese cabbage (CC), and leaf mustard (LM) on highland, Inthanon mountain.

Crop	Kind of Land	Seed Yield (kg/ha)	Income	Cost	Net Profit
			US$/ha		
CR	Slope	665.0	1259.0	905.4	353.6
CR	Paddy	456.0	877.4	468.6	408.8
LM	Paddy	169.0	324.5	572.6	-248.1
CC	Slope	498.8	959.1	882.3	76.8

Table 2. Percentage of input in the cost of production of Chinese radish seeds on highland, Inthanon mountain.

Item	Kind of Land	
	Slope (%)	Paddy (%)
Cost production	100.0	100.0
- Labour	55.6	48.3
- Fertiliser	23.2	35.4
- Insecticide	8.6	12.9
- Fungicide	8.3	2.3
- Others	4.3	0.7

Hill physiography, small land resource and harsh environment affect the distribution of fields or agricultural plots of individual farmers. Generally, Karen and Hmong cultivate several small scattered plots on the slopes and in the valleys. An individual may own more than five plots that may take more than one-day walking from his/her village. In recent years, farmers with vehicles travel longer distances. Due to extensive road construction and improvement during the last ten years, none of the villages in Inthanon is completely isolated.

The Hmong village of Khun Klang is clearly larger than a typical Karen village. However, there is a slight difference in average household size between both ethnic groups (Table 3). Culturally, the Karen family is a nuclear-structured form while Hmong family is an extended type. Studies indicate that the family form has an effect on management of labour use for agriculture. A system that requires a large amount of labour during the peak period of labour use is likely to limit chances for the Karen.

There are more than five projects and agencies' schemes in Khun Klang and Pa Mon, but only one or two agencies working in Nong Lom, Mae Klang Luang, and Sob Had. Farmers in these distant villages have less access to development support from both government and private agencies. Seed production extension has focused more on Khun Klang and Pa Mon.

Table 3. Population in Karen and Hmong Village

Village	Ethnic background	Number of household	Number of population	Average household size
Pa Mon	Karen	55	302	5.5
Nong Lom	Karen	49	239	4.9
Mae Klang Luang	Karen	31	184	6.0
Sob Had	Karen	14	73	5.2
Khun Klang	Hmong	130	799	6.2

Karen and Hmong are presently engaged in different production systems. They do not share economic pressures or agricultural choices. Rice is the main wet-season crop of the Karen while cabbage growing is a year-round cropping activity of the Hmong (Tan Kim Yong 1987). As compared to Hmong, who generate large cash income and use the large labour pool of their extended family, the Karen is a small-resource farmer who has to seek opportunity from wage employment. The Hmong cabbage cultivation uses hired labour from Karen villages. Based on field data, there is some evidence to confirm that most of Karen farmers tend to practice a diversified agriculture. A strategy of diversity is a way of responding to uncertainty about hazards and opportunities by spreading risk and expanding alternatives.

Both Karen and Hmong have an experience in forming several types of cooperative groups responding to development schemes, and in organizing indigenous groups for community activities. During the last five years, extensive activities of the Royal Project encouraged farmers to join the groups for vegetable growing, flower and fruit gardening, health and education program, credit, and a rice bank, etc.

Clearly, technical knowledge and skills of both Karen and Hmong have been improved through extensive training and extension of highland development schemes in Inthanon areas. As a result of adopting cabbage cultivation as their main production, the Hmong demonstrate more application of knowledge and practice new cropping more than the Karen. Technologically, both vegetable production and vegetable seed production is alike, though the degree of sophistication is different. To be successful to carry out various tasks in a new cropping system, it often requires efficient management of the inputs, timing and measurement, and marketing of the products.

SEED PRODUCTION ADOPTION AND PRACTICE

Seed production technology was introduced to Karen and Hmong farmers operating under different physical, economic and sociocultural conditions, Adopting such innovation obviously required experimentation, field practice and observation, and most of all, positive and active response from farmers. It was necessary for the technological transfer to receive sociological analysis, not just to determine current conditions but to refocus project staff onto practical problems and sociocultural aspects.

The purpose here is to illustrate the differences and similarities of adoption between Karen and Hmong. Eight Karen farmers and eight Hmong farmers interviewed were interested and accepted seed production at early stage of introduction. Both groups can be considered as early adopters when accounting for a chronological continuum of innovativeness as established by Rogers (1962). Through the adoption process, Karen and Hmong farmers took a short time to decide, get the production inputs, and apply new technology into their fields. Of three kinds of vegetable seed introduced, Chinese cabbage, leaf mustard and Chinese radish, only Chinese radish was accepted. Some of the small-resourced Karen farmers/adopters reported an economic setback, since in addition to losing an opportunity to earn extra income from wage employment during dry season, they also lost their seed crops due to technical and management problems.

New concepts and approaches in agricultural development require that farmers learn new techniques by experiencing them. Farmers should be put into positions to take risks on their own or to watch new techniques being applied, preferably by interacting with those who introduce them. In many areas this has been achieved by conducting experimentation on-farm. In this way, farmers and extensionists learned to act together in devising solutions to mutually defined problems.

The question: what are incentives for adoption? is important, but not as a focus in this present study. Based on the field data, the farmers' response to the questions for adoption showed that expected high profit margin was one of the major reasons. However, these simplistic views were not adequate to explain the complex nature of adoption decisions. Better understanding needs further sociological research which can indicate causal relationships of key variables.

Hmong farmers were more entrepreneurial than Karen farmers, and generally operated larger farm sizes. Having previous experience in the production of the commercial opium crop, Hmong farmers had the confidence to manage cash, supplies, and the product sale. Contact with the market and external agents to get information and other assistance is an important element in farmers' decisions. During the fieldwork, it was observed that Hmong farmers were not reluctant to approach the agents to seek for help. Major cropping of cabbage has generated large cash income for Hmong farmers throughout the year. This allowed them to be more flexible in managing hired labour for all types of production. Hired labour can assist the family during the peak period of labour demand. This was the reason why Hmong farmers were able to carry out seed production successfully even though it demanded a high labour cost. The report showed that labour used for production of Chinese radish seeds on slope land was about 52.4 day/rai. But to produce Chinese radish seeds on paddy field, Karen farmers used only less than 30 days/rai (Seed Production Project, 1987). In the Karen's practice of seed production, it was clear that the family was the only source of labour for all agricultural activities. Traditionally, most Karen farmers have to depend on an arrangement of labour exchange during the peak period of farm activities. A general observation from many peasant villages indicated that there is a tendency that this institution of labour exchange is rapidly eroding as more peasants are involved in commercial production. Conditions in family labour may be a constraining factor for seed production among Karen.

Physical location and distance of the village from the station, demonstration plot and from the extension service point explain the accessibility of seed growers to source of technical knowledge, production supplies, and information. All Hmong adopters interviewed live in Khun Klang where the station and extension service point are located. Hmong farmers could observe and learn from a demonstration plot, and approach the agents when they were on the sites. Pa Mon, the Karen village was selected to be the second service point. Two Karen adopters living in Nong Lom and Sob Had had to walk between ten to twenty kilometers to the site. This may reflect a low level of motivation on the side of Karen and a high risk of technical deficiency in their seed production. Almost no seeds were obtained from Karen farmers in the past years. Actually, Karen farmers were trained in the demonstration plots while Hmong farmers were not trained.

The cultivated areas in Inthanon are dispersed among several villages along a 40-kilometers stretch of the mountain region. Generally, an individual farmer manages to operate several small scattered plots. Some of the farmers have to spend a long week in the field shelter away from home village. When close to the harvesting time, some farmers may have to live there through the month.

One Karen adopter had to walk more than ten kilometers to the seed production field. Distance and location of plot may create less pressure on Hmong. Seed production was operated on slope land under rainfed or sprinkler irrigation. Paddy field location is generally where the canal irrigation is available. In addition, the Hmong farmers of Khun Klang own trucks and motorcycles which allow them to travel longer distance. They do not have to spent long periods at the field shelter, and therefore do not perceive social constraints to seed production.

Knowledge of the existing routine activities of wet season and dry season is crucial for the seed production staff in order to be able to plan the activities that fit farmer's schedule and needs. Traditionally, after rice harvesting, Karen are busily engaged in several economic and social activities. For Karen, dry-season (December-April) is the time for working as hired labour, raising buffalo of the lowlanders, collecting fuelwood, hunting, fixing house, etc. When the forest land was available in the old days, Karen usually explored a potential for clearing new cultivated fields during dry season. Especially, a newly married couple had to hunt for new fields. An important social activity is a search for the bride from the distant villages. These activities also occupied dry-season time. When a new technology or production system is introduced for operating during dry-season, an individual farmer has to calculate his or her opportunity cost. Clearly, an economic incentive is not always the interest of the Karen. Knowledge of local culture is obviously crucial to achieve this technological transfer.

Changing to seed production is a big shift in economic and technological practice of the Karen. They need close supervision and are more dependent on assistance of the external agencies at initial stage of adoption. The Karen's seed production practice also requires full package of inputs, credit, supervision, and marketing. Hmong farmers, who are better off as compared to Karen, have a tendency to need lower degree of the external support in inputs, credit, and supervision. These Hmong farmers have a high potential of being more independent.

The study indicates that Karen and Hmong need different degrees of extension services. Those who have high potential and low pressure such as Hmong farmers at Inthanon should receive less services. The seed production project should plan to provide a full assistance package, and high degree of extension supervision, to Karen. Many agricultural development schemes implemented in different countries have failed because the planners simply assumed a uniform pattern of extension for all target groups.

In actual operation during 1986 to 1988, the seed production project of CMU has implemented an extension program which was obviously of more benefit to Hmong farmers who were close to the service points and station. Karen adopters who clearly needed more assistance actually received less extension services due to distance and location of their villages and their fields (Tables 4 and 5). Under local conditions, this type of extension demands a continuing, consistent effort from extensionists, and has a high economic cost.

Table 4: Seed Production Extension 1987

	Introduced by		Technical Skill	Visit Demonstration Plot	Seed	Fertilizer Visit	Extensionist	Marketing	Extra Extention Service Of Other Agency		
Adopter	Relative	Extentionist							Royal Projects	Public Welfare Department	Others
Hmong											
Case I	X	-	-	-	X	-	-	X	-	-	-
Case II	X	X	X	-	X	-	-	X	X	-	-
Case III	-	-	-	-	-	-	-	-	-	-	-
Case VI	X	X	X	-	X	-	-	X	X	-	-
Case V	X	-	-	-	X	-	-	X	Z	-	X
Case VI	X	X	X	-	X	-	-	X	X	-	-
Case VII	X	-	-	-	X	-	-	X	X	-	X
Case VIII	X	X	X	-	X	X	-	X	X	-	-
Karen											
Case I	-	X	X	X	X	X	-	-	X	-	-
Case II	-	X	X	X	X	X	-	-	-	X	-
Case III	-	X	X	X	X	X	-	X	-	X	-
Case VI											
Case V											
Case VI	-	X	X	X	X	X	-	X	-	X	-
Case VII	X	-	-	-	X	X	-	X	X	-	-
Case VIII	-	X	X	-	X	X	-	-	X	-	-

Table 4: Seed Production Extension 1987

Adopter	Introduced by		Technical Skill	Visit Demonstration Plot	Seed	Fertilizer Visit	Extensionist	Marketing	Extra Extention Service Of Other Agency		
	Relative	Extentionist							Royal Projects	Public Welfare Department	Others
Hmong											
Case I	X	-	-	-	X	-	-	X	-	-	-
Case II	X	X	-	-	X	-	-	X	X	-	-
Case III	X	-	-	-	X	-	-	X	X	-	-
Case VI	X	X	X	-	X	-	-	X	X	-	-
Case V											
Case VI											
Case VII											
Case VIII	X	-	-	-	X	-	-	X	X	-	-
Karen											
Case I	-	-	-	-	-	-	-	-	-	-	-
Case II	-	-	-	-	-	-	-	-	-	-	-
Case III	-	-	-	-	X	X	-	-	-	X	-
Case VI	X	-	-	-	X	X	-	-	-	-	-
Case V	X	-	-	-	X	X	-	-	X	-	-
Case VI	-	-	-	-	X	X	-	-	-	X	-
Case VII	X	-	-	-	X	X	-	-	-	X	-
Case VIII	X	-	-	-	X	X	-	X	-	X	-

FARMER'S PERCEPTION ON PROGRAM AND PROBLEMS

What the farmers or adopters think about a program is very important for the project management to improve or to adjust an extension strategy and farmer's production. All farmers interviewed think that the seed production is useful and can be one of their economic alternatives. The seed production when efficiently managed can bring extra income to their families. The project scientists should seek ways to reduce the farmer's risk in production and marketing. From the interview, most of the farmers expressed their wish to continue to produce seed in the following years, and were anxious to learn more.

Non-adopters interviewed also indicated a positive response to the seed production program. However, some of them were reluctant to accept because they did not have sufficient labour. Other reasons for a delayed adoption were an insecurity of land tenure and lack of opportunity to interact directly with the project staff or the extensionists.

From the project document and the interview of the field personnel, the extension works were carried out according to the guidelines. The procedural steps to work with farmers were as follows:

1. To introduce seed production to farmers, and encourage them to learn from the demonstration plot.

2. To provide technical and economic information for the interested farmers, and increase their attention by person-to-person visits.

3. To teach farmers the techniques, steps, and tools for seed production and marketing.

4. To distribute seeds and production supplies to the adopters who would now be considered as contract farmers under the program.

5. To provide supervision and monitor the production at the sites, and help to solve problems.

6. To teach farmers the harvesting techniques, post-harvesting methods, and marketing.

7. To provide assistance on marketing by buying all seed products from the adopters.

The data obtained from the field showed that some of the adopters firstly learned about the seed production from their relatives and the progressive farmers of their villages. Kinship was likely to be useful to support this technological transfer. This cultural process was functioning actively in both villages of Hmong and Karen. Some of the adopters were introduced directly to the extensionists. Especially, farmers of Khun Klang and Pa Mon had more opportunity to interact directly with the project staff. Actually, sixteen adopters received different degrees of extension services. Services provided for Karen adopters were partial, less intensive, and less frequent. The strategy of extension services for both Karen and Hmong should be redirected for future implementation.

Perceived problems of adopters were recorded as related to irrigation, drought, insect activity, distance of the seed production plots, family labour supply, marketing and price, and extension services. Some of these problems can be solved by the technical solutions in seed production technology. Some of them will be eliminated when the project manager and staff apply new tools and procedures to identify potential farmers and appropriate location of plots.

Extension problems perceived by farmers/adopters were low frequency of visit, delay of inputs delivery system and on-site supervision. Loss from drought and insect activity perhaps could be reduced if farmers have had a chance to consult with the extensionists. Although farmers were very satisfied with the available full package of assistance, including inputs, technical knowledge, and marketing, some of them had less access to services. These people were the farmers who expressed the problems mentioned.

RECOMMENDATIONS

The study indicates a variation in adoption and practice of seed production among ethnic communities of the hill region of northern Thailand. A process of experimentation, technological innovation, and transfer clearly involves a complex interaction of technological, economic, and sociocultural elements in a seed production program. This requires teamwork of multi-disciplinary science and social sciences. This present study can only provide some of the important aspects of adoption behavior and pattern. Better understanding of perceived problems of farmers is useful to practical works of the seed production project.

Here, it is appropriate for this study to present some recommendations.

1. To reduce risk in technical deficiency and inappropriate management, the project should carefully plan to work with the potential, selected farmers, and to set criteria for selecting target farmers at early stage.

2. To avoid the issues of location and distance of seed production plots, mutual decisions of farmers and extensionists should be encouraged to determine potential sites.

3. To establish a strategy and approach for provision of different degree of extension services.

4. To establish both partial assistance and full assistance in the extension program to fit different groups of adopters.

5. To seek better understanding of sociocultural characteristics of local communities to improve knowledge and skill of the project staff in working with farmers.

6. To plan for and create qualified extension personnel who have incentives and commitment to work with farmers by establishing an appropriate system of recruitment, training and rewards.

7. To redirect the project into the group approach, not to emphasize on individual farmer approach, in order to increase efficiency and reduce operating costs, when extensive seed production is to be implemented in several provinces of Northern Thailand.

SOURCE OF INFORMATION

Carter, William B. and Edwin P. Bugbce, Jr., The Marketing of Seeds, Yearbook of Agriculture, 1961, pp. 492-499.

Department of Agriculture Extension, Statistics on Vegetable Product, Ministry of Agriculture, Thailand, 1986/87.

Faculty of Agriculture, Vegetable Seed Production, Chiang Mai University, Chiang Mai 1986 and 1987.

FAO Walther P. Feistritzer (ed.) Improved Seed Production: A Manual on the Formulation, Implementation and Evaluation of Seed Programmes and Project. Rome. Plant Production and Protection Series No. 15, 1978.

Kamol Lertratana, Data on Vegetable Seeds in Thailand, Kan Kaset, No. 13, Vol. 6 (November-December), 1985.

Office of Commerce, Data on Trade, Chiang Mai, 1986.

Office of Commerce, Data on Trade, Chiang Mai, 1987.

Pithoon Rodvinji, Seeds: Price Policy and Marketing, MOAC, 1984.

Rogers, Everett M. Diffusion of Innovations. New York: The Free Press, 1962.

Rollin, S.F. and Frederick A. Johnson, Our Laws That Pertain To Seeds, Yearbook of Agriculture, 1961, pp. 482-492.

Tan-Kim-Yong, Uraivan and Sanay Yarnasarn, et.al. Resource Utilization and Management in Mae Khan Basin: An Intermediate-Zone Crisis, Chiang Mai University, Chiang Mai, 1987.

CHICKPEA AND LENTIL SEED PRODUCTION AND DISTRIBUTION MECHANISMS IN PAKISTAN

A.M. Haqqani (S.S.O.) and M. Riaz Malik (S.O.)
Pulses Program, National Agricultural Research Centre
Islamabad

SUMMARY

Chickpea is one of the most important pulse crops, grown annually on about 1.0 million hectares with a production of 0.5 million tonnes. 85-90% of the crop is primarily grown under rainfed conditions of the Punjab and N.W.F.P. Provinces. A survey was conducted from 21st February - 9th March, 1990 in the major chickpea production zones and lentil major production district of Sialkot of the Punjab and N.W.F.P. Provinces with the objectives:

(i) to study the seed production and dissemination systems of improved chickpea and lentil cultivars;
(ii) to know the socio-economic systems of chickpea and lentil growers;
(iii) to estimate the amount of improved seed distributed and number of growers using it with special reference to resource poor farmers; and,
(iv) to estimate the potential production of chickpea and lentil keeping in view the crop condition in relation to diseases and pest infestation.

Research institutes are the pioneers in producing the pre-basic seed, and after varietal evaluation, it is handed over to the Punjab Seed Supply Corporation (PSSC), which distributes it to the government/private farms for seed multiplication purposes. For maintenance of seed purity, agricultural scientists have been deputed to visit the seed multiplication farms at least 2-3 times a year. After seed multiplication, improved seed is returned to PSSC, which sends this seed to its sale points for sale to farmers. Seed is mostly sold by the market agents (Arthis), who usually manipulate the price at the time of sowing and harvesting and thus paralyse the price mechanism/marketing system.

In main growing areas, chickpea-fallow-chickpea is the dominating cropping pattern, but in the areas where some water is available cropping patterns are as follows: wheat-fallow-wheat, chickpea-groundnut-chickpea, lentil-rice-lentil. About 70%, 68%, 62%, 83% and 76%, of the farmers are adopting improved technology in terms of ploughing and planking, recommended seed rate, use of improved seed, drill sowing and weeding respectively. 93% of the farmers are aware of the improved/ commercial varieties. 19% of the farmers are using credit, whereas 19% reported that credit is available but that they don't use it because procedures for loan and repayment are too cumbersome; 62% complained that there are social constraints to approaching credit institutions.

The current survey showed that crop condition as expressed by the farmers is relatively satisfactory compared with previous 2-3 years, when the growing areas experienced a long drought spell, which curtailed the production level from 486,000 tonnes to 371,000 tonnes.

INTRODUCTION

The International Development Research Centre gave approval to the Pakistan Agricultural Research Council for the Food Legumes (Pakistan) Project in 1979. A second phase began in 1985 and ran until 1989. Several institutions were involved in this Project:

Ayub Agricultural Research Centre, Faisalbad
University of Agriculture, Faisalbad
Agricultural Research Station, Dhudial, Mansehra
Rice Research Institute, Dokri, Larkana
Agricultural Research Institute, Sariab, Quetta

The coordinating Centre, located at the National Agricultural Research Centre (NARC) in Islamabad, carries out the following functions:

a) Collection, exchange, screening and distribution of germplasm and breeding material from abroad and within Pakistan. This material was shared with all the provincial units in the country.
b) Liaison with International Centres (ICRISAT, ICARDA, AVRDC), and with pulses projects in other countries in the region.
c) Organization of pulses seminars, workshops and study tours, and provision of literature.
d) Arranging local and foreign training of food legume scientists, and consultancies when required.
e) Administration of project funds to the provincial centres, and reporting of coordinated research.

In Pakistan, varieties are being developed through conventional plant breeding methods, i.e. by inducing and creating genetic variability through crossing and hybridization, and thereafter exerting selection pressure. In addition to this, the Research Institutes receive every year thousands of lines and other germplasm of wider genetic character developed by the International Centres. This material is screened under diverse environmental conditions. Characters such as distinctness of form, uniformity, stability, resistance to pests and diseases, and other agronomic traits contribute to the superiority of a variety.

PROJECT RESULTS

By objective, the results have been as follows:

Objective a) To evaluate pulse germplasm and breeding lines for yield, disease resistance, and other characteristics of economic importance.

Mungbean (600 lines), mash bean (1000) and lentil (650) were evaluated. Forty mungbean lines and seventy-six lentil lines showing good performance on the bases of yield per plant, days to 50% flowering, lodging resistance, disease resistance and high yield potential were selected for further study. From the mash germplasm, 730 single plants were selected on the basis of early maturity, tolerance to diseases and high yield potential for further evaluation in 1986-87. In 1987-88, these single plant progenies and 750 local germplasm accessions were evaluated, and 156 lines were selected on the basis of good plant type, early maturity and other traits. Two groups of material in mash (V. mungo) have been developed - the early 60-day crop for two crops per year, and the normal summer planting type.

At the University of Agriculture, Faisalbad, 141 entries of lentil germplasm (54 from NARC, 17 from the Department of PBG, AUF, Faisalbad, and 70 from the pulses section, AARI, Faisalbad) were evaluated against A. lentil in the greenhouse and field. The entries which exhibited resistant reaction to A. lentil in the greehouse were ILL-4605, 33831, 33121, E8-8, E-11-13, E-4-1, E-8-6, E-11-12, E-8-10 and E-17-1, while the entries which exhibited resistance reaction in the field were ILL-4605, 33661, 33704, 33704-1, 85-5-33201, E206, E-4-1, E-8-6 and E-11-12.

At NARC, Islamabad, 196 lentil accessions of local and exotic origin were evaluated against collar rot under natural field conditions. Out of 196 lines, 102 were found free of disease, 32 entries were resistant, and 13 were tolerant, while others were susceptible or highly susceptible to this disease.

Objective b, i) Through breeding and selection to develop high-yielding and stable yielding varieties of lentil, mungbean and black gram.

Lentil. The selected material from germplasm and entries obtained from abroad were evaluated in five preliminary and four major varietal trials. The check variety `precoz' was early in flowering and maturity, while 9-6 outyielded all the other varieties, followed by 785 26004 x Pant L 406-1, 74 TA 441 x Pant I 639, Giza x ILL-1, L-9-12, 74 TA 938 and L-912 x 76 TA 6654. In National Uniform Yield Trials conducted at 12 locations in the country, AARIL-334, AARIL-502, AARIL-498, and AARIL-337 showed good performance during 1986-87 and 1987-88. If the project is further extended, the top-yielders will be tested on farmers fields. `Precoz' has been released in the name of Masehra-89 for commercial cultivation by Research Station Dhudial, Mansehra.

Mung. During 1986-87 and 1987-88, preliminary, major and National Uniform Yield Trials were conducted at NARC Islamabad, AARI Faisalbad, University of Agriculture Faisalbad, and ARS Sariab Quetta. The varieties NHM-54, BRM-114, NCM-7, NHM-45, NHM-51 and NCM-69 were higher yielders and showed yield stability.

Mash. AARI Faisalbad conducted a yield-performance trial, and found that Quandari Mash out-yielded the other cultivars giving 1.5 t/ha. The check variety mash-48 gave the yield of 769 kg/ha. In 1987-88, in another trial conducted under different ecological zones, AARIM-118 out-yielded other cultivars (793 kg/ha) on an average of five locations, while at Khanpur location it gave a yield of 1597 kg/ha. The other promising cultivars were mash-59, mash-216 and AARIM-13.

Objective b, ii) Develop early maturing cultivars of lentil and chickpea for late planting in rice-based cropping systems.

Lentil. Precoz, a bold-seeded lentil, has been identified as being resistant to blight and rust, and is also very fast growing (offering better competition to weeds) and early maturing. The other three early maturing lentil lines are AARIL-337, Flip 86-38L and AARIL-502.

Chickpea. Breeding material (F2-F7) developed by NARC Islamabad was planted in rice fields in 1986 to select the material suitable for cultivation after IRRI (semidwarf) and basmati (local tall aromatic) varieties. Four crosses (F4-F6) PK-51825xCM-72, PK-51832xCM-72, PK-51835xCM-72 and PK-51863xNEC-138-2 were selected as resistant to blight and suitable for cultivation in a post-rice environment. Single-plant selections were also made from these crosses for further testing.

Mungbean. Out of 32 promising lines screened against mungbean yellow mosaic virus (MYMV) in the field, 13 entries showed resistance to this disease. The same lines were tested against Cercospora leaf spot (CLS) and 16 were recorded as resistant (E-321, NCM-69 and NEC-68 showed multiple resistance against MYMV and CLS). 1988 was a disease-free year and screening could not be performed effectively under natural conditions.

The same 32 lines were tested against MYMV at the University of Agriculture, Faisalbad, during 1986-87 and 1987-88. Eight entries showed moderate resistance to MYMV, while the remainder were highly susceptible.

Lentil. Four nurseries (local and exotic) were evaluated at NARC for disease resistance against Ascochyta blight, botrytis, stem rot and anthracnose diseases. Botrytis and stem rot were very common. The lines ILL-5527x113458, lairdxprecoz, ILL-5732, ILL-1939, ILL-5585 and LG-116 were noted as tolerant to stem rot. In the lentil National Uniform Yield Trial, two lines LP-1 and LP-2 were free of Ascochyta lentis and botrytis, while precoz was resistant to blight.

Out of 196 germplasm lines, a couple were found free of collar rot, 32 entries were resistant and 13 were tolerant, while others were susceptible to highly susceptible. From a lentil blight screening nursery at ICARDA, 11 lines were resistant, 15 tolerant and 5 susceptible

Objective c) To develop appropriate production technology suitable for food legumes in rainfed areas and irrigated conditions.

Mungbean and mashbean. The research on the effect of rhizobium inoculation on four planting arrangements showed that the triple row stripsowing inoculated treatment was more productive as compared to others while in mashbean, the single row inoculated treatment was more productive.

In the corn-mungbean intercropping trial, maximum grain yield of the main crop as well as the component crop was recorded when the corn was sown in double row strip mungbean. In a chemical weed control trial, fusilate in combination with flex gave very promising results in two years testing in both mungbean and mashbean crops.

Chickpea. Research on sowing date x genotype showed that mid-November is the best sowing date in Islamabad and Rawalpindi regions as far as yield is concerned. In a spacing trial, 10 x 30 cm proved to be the best.

Lentil. Four sowing dates with three genotypes were tested, and the optimum date was marked as 16 October, with the latest date for crop maturity being 1 November. In a spacing study, 25cm rows proved the optimum for grain yield.

Objective d) To test and refine pulse technology through on-farm testing.

On-farm trials on chickpea, lentil and mungbean were conducted in the Fatehjang area in combination with farming systems research at NARC. The results are encouraging, but are not reported here.

Objective e) To develop the capability of raising off-season crops of chickpea, lentil, mungbean and black gram for seed increase and generation advance at high elevation locations.

Twenty seven F1s, varying numbers of single-plant progenies of F4s, and F4 bulks were grown at the high altitude research station at Kaghan, to advance a generation. Five elite lines of advanced crosses were also grown to increase seed. These populations produced F2 seed successfully, which was then planted during the winter season at Islamabad. Fresh crosses were also made at Kaghan, which have also been planted in an NARC breeding block.

SEED PRODUCTION AND DISSEMINATION SYSTEMS

The complete structure of chickpea and lentil seed production and dissemination system has been portrayed in Figure 1. Chickpea and lentil cultivars are being developed by breeders at their research institutes. Pre-basic seed is produced during varietal development by breeders and after registration by the National Seed Registration Department and approval of Varietal Evaluation Committee the basic seed is handed over to Punjab Seed Supply Corporation (PSSC) for multiplication.

VARIETAL RELEASE POLICIES

There are three agricultural universities, 12 multidisciplinary institutes and 31 monocrop research institutes in the country that work on varietal development for local agroecological conditions. The procedures for variety release are as follows:

Testing. The breeder includes promising lines in Zonal Varietal Trials. These trials are conducted in cooperation with selected farmers and at Government Farms. Lines that perform better than the locally-adopted commercial check variety are then tested in bigger blocks on private and Government farms. After final testing, the National Seed Registration Department (NSRD) and the National Seed Council (NSC) release the varieties for commercial cultivation.

NSC has the full authority for policy formulation, and for setting up and regulating the production and quality of seed. It represents all the disciplines concerned in the public and private-sector Seed Industry. NSRD, as the executive arm of the NSC, performs the following functions:

- i) Pre-registration of varieties, for the purpose of determining suitability for registration as a variety, providing definitive botanical descriptions and information on genetic suitability and adaptability.
- ii) Registration of varieties.
- iii) Publishing a list of registered varieties.
- iv) Performing other functions as assigned by NSC.

Procedure for registration and approval. In order to regulate and coordinate the functions of research institutes and other organizations, and to evaluate and release the varieties, the following procedure was adopted by the NSC:

- i) When the breeder thinks that a selection is nearing the final stages of testing, he sends the plant material simultaneously to PARC (or to PCCC for cotton) for testing disease reaction, agronomic traits, performance and suitability, and to the NSRD for testing for morphological characters.
- ii) The Variety Evaluation Committee in the PARC continues to examine varieties for the above traits, and reports its findings to the Federal Registration Committee.
- iii) PARC and PCCC conduct their tests and inform the breeders and the

Figure 1: Complete Structure of Seed Production and Distribution Mechanism

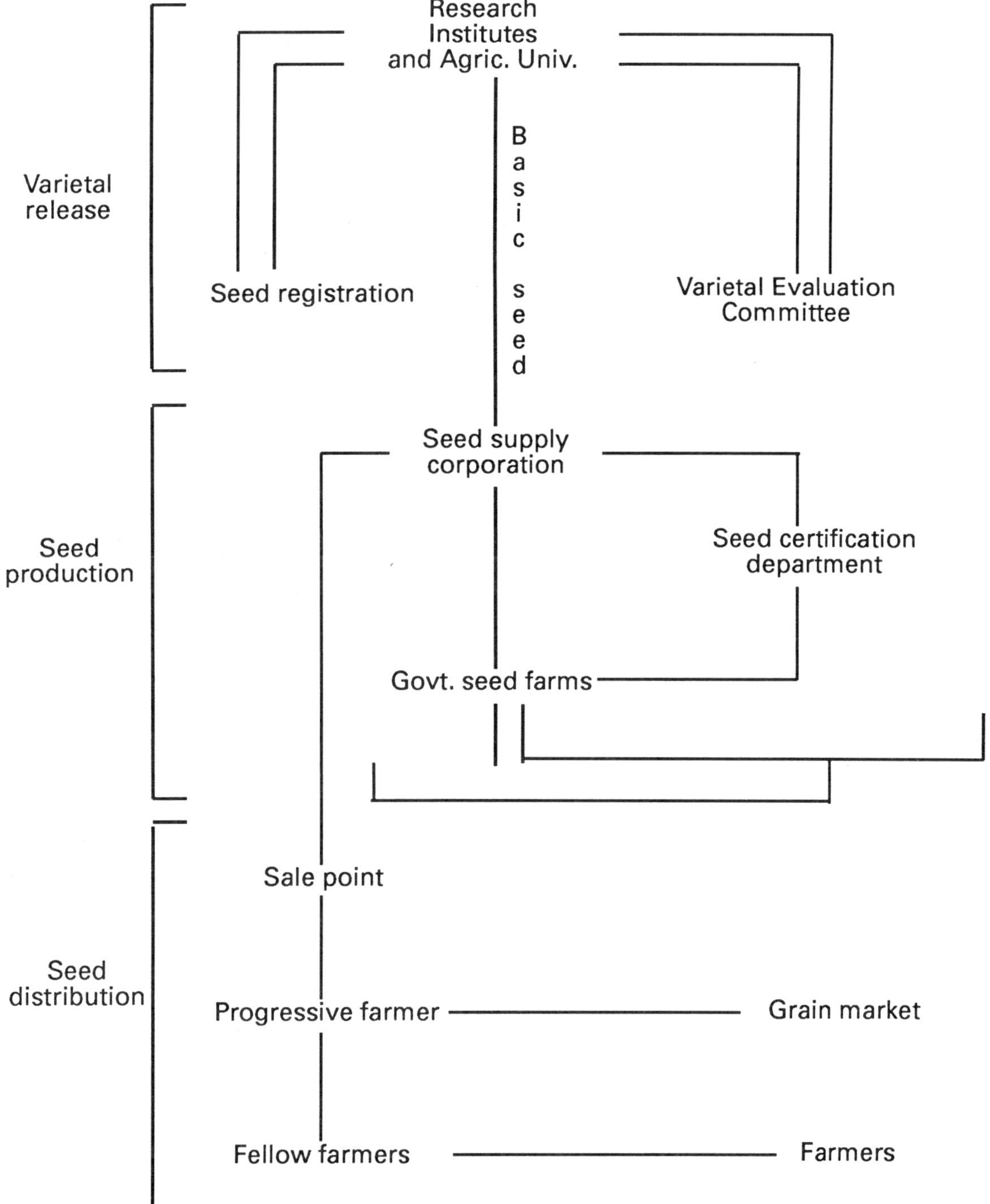

Provincial Governments.

iv) The Provincial Variety Approval Committee examines the breeders data, as well as the report from PARC/PCCC/NSRD, before approving and recommending any variety. The NSRD submits the registration reports of the Federal Seed Registration Committee to NSC for decisions on registration and release.

v) NSRD independently checks distinctness of a variety, and puts its findings to the Registration Committee.

vi) The NSC delegates the Provincial Seed Councils (PSC) to release those varieties on which the PSC participants reach consensus. The PSC considers and evaluates data from provincial breeders and agronomists,as well as that from the tests and trials conducted by the NSRD and PARC Varietal Evaluation Committee through the National Uniform Yield Trials, in which all breeders from the country provide their promising cultivars.

vii)In case of a difference of opinion in the PSC about the approval of a certain variety, it was decided by the NSC that the matter would be submitted to NSC through the Seed Registration Committee, which would incorporate the views of the members of the council for final decision by NSC. Afterwards, the varieties approved by the NSC would be notified for general release.

Constraints.

i) Paucity of budget for conducting multilocational testing of candidate varieties.

ii) Breeders are afraid of registering their varieties with the NSRD due to lack of confidence regarding secrecy of the material.

iii) The NSRD does not have the equipment and chemicals used in chromosomal mapping, necessary for authenticating ownership of morphologically-alike varieties.

PUNJAB SEED SUPPLY CORPORATION

PSSC sends the seeds to Government Seed Farms located in major chickpea and lentil production zones where seed is multiplied. Table 1 presents the details of Government Seed Farms (GSF) visited. From this table it can be seen that land for improved seed production of chickpea and lentil is much less than required to meet the demands of farmers in their respective districts.

For the maintenance of purity of seed, agricultural scientists visit the GSF three times each season. The first visit is conducted at flowering, second at pod formation and third at maturity. They instruct the farm manager to eradicate off-type plants and to apply insecticides and fungicides to control diseases and insects, if any. The main objective of these visits is to maintain the seed purity.

The chickpea and lentil seed of GSF is certified by the seed certification department and PSSC is allowed to procure the seed from the GSF. PSSC has sale points at the cities and towns accessible to farmers. Farmers buy the improved seed from the sale points of PSSC in Punjab, and of Agricultural Development Agencies in NWFP.

GRAIN MARKETS

The Arithis (Sale Agent) procure improved chickpea and lentil cultivars from the progressive farmers. At the time of growing season farmers buy the seed from the grain market but the authenticity and purity of the seed is not guaranteed. The Sale Agents procure the seed at normal rates from the farmers and sell it at much higher rate at the time of sowing, increasing the production cost. The resource-poor farmers cannot buy the seed at higher rates and are bound to sow their aboriginal cultivars.

FELLOW FARMERS

The farmers also get the improved seed from their fellow farmers who have access to GSF. This method of seed distribution is of vital importance and plays a key role in improved seed spread among resource-poor farmers particularly.

ON-FARM SURVEY

In Pakistan, chickpea, lentil, green and black gram are the major food legumes, cultivated on about 1.5m ha. Chickpea is the most important crop, being grown on about 1m ha annually, with a production of about 0.5m t. 85-90% of the crop is primarily grown under rainfed conditions in the Punjab and NWFP Provinces. In the main areas, chickpea-fallow-chickpea is the dominant cropping pattern, but in the areas where some water is available, cropping may follow wheat-fallow-wheat, chickpea-groundnut-chickpea, or lentil-rice-lentil.

Thal region (Bhakkar, Khushab, Layyah and Mianwali) was surveyed to study the seed production and distribution mechanisms where low-income farmers produce subsistence and cash crops with minimal dependence on external inputs. For lentil, Sialkot district was explored thoroughly. In addition Chakwal, Tala gang, D.I. Khan, Bannu and D.G. Khan were also surveyed. The farmers who achieved moderate to high level of chickpea production were interviewed. The system of seed production and seed distribution at Government farms and sale points (Table 1) were studied. The objectives of this survey were:

1. To study seed production and dissemination systems of improved chickpea and lentil cultivars.
2. To determine the socio-economic conditions of the chickpea and lentil growers.
3. To estimate the amount of improved seed distributed and the number of growers using it, with special reference to resource-poor farmers.
4. To estimate the potential production of chickpea and lentil, with reference to crop condition and pest outbreaks.

SURVEY FINDINGS

Out of 103 farmers interviewed, 59 were small farmers (up to 10 ha), 18 medium (10-20 ha) and 26 large farmers (above 20 ha). The cropping patterns adopted by these farmers are presented in Table 2. Overall, 41% farmers were using improved seed of chickpea and lentil, while the rest of the farmers were using their local seed (Table 3). The farmers who were not using improved seed were aware of commercial varieties released by the Agricultural Research Institutes and Universities but they did not know the sources of improved seed. These farmers complained that no extension services are being rendered to them for their guidance. There were only 10% farmers who were not aware of the modern chickpea and lentil production technology but they were eager to know recent developments made in crop husbandry of these crops.

Credit facilities are one of the most limiting factors in adoption of new production technology as there were only 19% farmers who were receiving credit from the Government Financial Institutions. Sixty one percent farmers complained that credit is needed but not available to them because of lengthy and complex procedures of getting the credit. There were only 19% farmers who claimed that they are not in need of credit (Table 4). Because of illiteracy and unawareness of the complex procedure, the small farmers are hesitant to approach the financial institutions. The major benefit of these facilities is enjoyed by the big landlords, which is a great set back for the dissemination of production technology. There were also 6% big landlords who were not using improved technology because of their illiteracy. There were more small farmers compared to big landlords, who were adopting new production technology (Table 5). This indicated that small farmers are interested to use the modern production technology but they are restrained by their resource-poorness and illiteracy.

CONSTRAINTS

The important constraints in augmenting chickpea and lentil production described by the growers in chronological order of importance have been categorised as follows:

a) Physical/Biological Constraints

i. Marginal land for chickpea and lentil production.
ii. Dissemination of improved seed specially resistant to Ascochyta blight and Fusarium Wilt in Chickpea and rust-free seed of lentil in major lentil growing areas.
iii. Weed problems.
iv. Insect problems.
v. Uncertainty of rainfall.
vi. Scattered and small land holdings that prevent efficient adoption of technology.

b) Technical Constraints

i. Poor threshing quality (5-10% seed breakage).

c) Economic Constraints

i. Lack of credit facilities particularly to resource-poor farmers.
ii. Unregulated marketing system.
iii. A big gap between producers gain and price paid by the consumer.
iv. Low incentive associated with share cropping and other land tenure systems.
v. High risk to apply agricultural inputs for higher yield.
vi. Inadequate extension services for dissemination and adoption of improved production technology.

POSSIBLE APPROACHES TO IMPROVEMENT

1. To improve farmer knowledge of new varieties and technology, there should be more farmer days, agricultural fairs and meetings with extension workers.

2. Credit procedures should be simplified.
3. Extension services should be strengthened through provision of budget

and infrastructure.

4. Farmers should be credited the price of improved seed at sowing time to maintain seed costs at a reasonable level.

CONCLUSIONS

The study suggested that more land and budget should be provided to Government Seed Farms for more lentil and chickpea seed production and PSSC should increase the sales points for better distribution of improved seed. Credit should be available to all categories of farmers on flexible terms and conditions for purchase of inputs and farm equipment. Strenuous research efforts are needed to combat the severe weed problem in the form of formulating pre and post weed emergence weedicides. Similarly, insecticides should be available in time at accessible places. Mansehra 89 is recommended for cultivation on large scale in Sialkot area. Moreover, modern production technology of chickpea and lentil must be disseminated actively by extension services before the sowing season starts.

Table 1: CHICKPEA PRODUCTION FARMS/SALES POINTS

A: Govt. Seed Farms	Area/ha	Production/Tonnes
1. Lentil Res.Station Sahu Wali, Sialkot.	8	10
2. Govt.Seed Production Farm Piplan (Liaquat Abad) Mianwali.	123	137
3. Atomic Energy Seed Production Farm, Kundian, Mianwali.	79	12
4. Gram Seed Production Farm Rakh Utra (Shah Wala) Johar Abad, Khushab.	194	48
5. Govt. Seed Production Farm Krore, Distt. Leiah.	8	5
6. Gram Res. Station Kallur Kot, Distt. Bhakkar.	14	9
7. Govt. Seed Farm Rakh Manghan D.I. Khan.	20	15
B: Private Farms		
8. Nadeem Model Farm	5	9
9. Mr. Abdul Aziz Malik Sales Agent (PSSC) Sialkot.	-	5
10. Mr. Mulazam Sales Agent (PSSC) Noor Pur Thal, Khushab.	-	260
11. M.Bashir & M.Amir Sales Agent (Pvt)	-	245

Table 2: DISTRICT-WISE DOMINATING CROPPING PATTERN

DISTRICT	CROPPING PATTERN		
Chakwal	Chickpea	- Mung	- Chickpea
	Wheat	- Maize	- Wheat
	Chickpea	- Sorghum	- Chickpea
	Lentil	- Mash	- Lentil
	Chickpea	- Groundnut	- Chickpea
Sialkot	Wheat	- Fallow	- Wheat
	Lentil	- Fallow	- Lentil
	Lentil	- Mung	- Lentil
	Lentil	- Mash	- Lentil
	Wheat	- Sorghum	- Wheat
	Rice	- Wheat	- Rice
	Rice	- Lentil	- Rice
	Lentil	- Maize	- Lentil
Attock	Chickpea	- Groundnut	- Chickpea
	Chickpea	- Sorghum	- Chickpea
	Chickpea	- Guara	- Chickpea
	Chickpea	- Mung	- Chickpea
	Chickpea	- Mash	- Chickpea
	Wheat	- Fallow	- Wheat
	Wheat	- Sorghum	- Wheat
D.G. Khan	Chickpea	- Guara	- Chickpea
	Chickpea	- Sorghum	- Chickpea
	Chickpea	- Rice	- Chickpea
	Wheat	- Rice	- Wheat
	Wheat	- Cotton	- Wheat
Leiah	Chickpea	- Fallow	- Chickpea
	Chickpea	- Guara	- Chickpea
	Wheat	- Fallow	- Wheat
	Wheat	- Guara	- Wheat
Bhakkar	Chickpea	- Fallow	- Chickpea
	Wheat	- Fallow	- Wheat
	Wheat	- Cotton	- Wheat
	Chickpea	- Guara	- Chickpea
	Wheat	- Guara	- Wheat
Khushab	Chickpea	- Fallow	- Chickpea
	Chickpea	- Guara	- Chickpea
	Chickpea	- Sorghum	- Chickpea
	Chickpea	- Mung	- Chickpea
	Chickpea	- Mash	- Chickpea
	Wheat	- Sorghum	- Wheat

Table 2: DISTRICT-WISE DOMINATING CROPPING PATTERN

DISTRICT	CROPPING PATTERN		
Mianwali	Chickpea	- Fallow	- Chickpea
	Chickpea	- Guara	- Chickpea
	Chickpea	- Water Mallon	- Chickpea
	Wheat	- Fallow	- Wheat
	Wheat	- Cotton	- Wheat
	Wheat	- Guara	- Wheat
Karak	Chickpea	- Fallow	- Chickpea
	Chickpea	- Sorghum	- Chickpea
	Chickpea	- Millet	- Chickpea
	Wheat	- Fallow	- Wheat
	Wheat	- Sorghum	- Wheat
	Wheat	- Millet	- Wheat
Bannu	Chickpea	- Fallow	- Chickpea
	Chickpea	- Millet	- Chickpea
	Chickpea	- Sorghum	- Chickpea
	Wheat	- Fallow	- Wheat
D.I. Khan	Chickpea	- Fallow	- Chickpea
	Chickpea	- Sorghum	- Chickpea
	Chickpea	- MIllet	- Chickpea
	Wheat	- Fallow	- Wheat

Table 3: VARIETAL ADOPTION

Farmer Status	No of Farmers	Farmers Using Improved Seed	Farmers Using Local Seed	Farmers aware of Improved Seed	Farmers unaware Improved Seed
Small (1-24 Acres)	59	15	44	49	10
Medium (26-50 Acres)	18	11	7	18	-
Large (50 Acres & above)	26	16	10	26	-
Total	103	42	61	93	10

Table 4: CREDIT AVAILABILITY

Farmer Status	Credit Received	Credit available but not needed	Credit not available
Small	3	8	48
Medium	5	2	11
Large	12	10	4
Total	20	20	63

Table 5-A: ADOPTION OF PRODUCTION TECHNOLOGY

Farmer Status	No. of Farmers Adopting Technology				
	Ploughing & Planking	Seed rate	Improved seed	Method of sowing (Drill)	Weeding
Small	37	38	32	45	46
Medium	14	12	13	15	12
Large	21	20	19	26	20
Total	72	68	64	86	78

Table 5-B: NON-ADOPTION OF PRODUCTION TECHNOLOGY

Farmer Status	No. of Farmers Not Adopting Production Technology				
	Ploughing & Planking	Seed rate	Improved seed	Method of sowing (B.casting)	Weeding
Small	22	21	27	14	13
Medium	3	5	4	2	5
Large	6	9	8	1	7
Total	31	35	39	17	25
Grand Total		103	103	103	103

SEED REVOLVING FUNDS: EXPERIENCES IN DISSEMINATION MECHANISMS IN PEASANT COMMUNITIES OF PERU

Faustino Ccama,
Adolfo Achata
and Francisco Torres.

PISA Project,
National Institute for Agricultural and Agroindustrial Research
Puno, Peru.

Abstract

Revolving funds (RFs) have been established by the PISA Project as a means of increasing capital availability to small farmers, to help them cope with climatic and economic constraints. While established as a service to the target communities, the funds also serve as a research tool, to test this mechanism before recommendiong it as a strategy for widespread use.

The funds provide a variety of inputs: seed, veterinary products, fertilizer, and sometimes cash. Initially four communities were targeted, and both community and personal loans were made. Individual recipients could specify whether or not they wished to receive technical assistance.

The funds have been in operation for up to four seasons. Seven communities are now covered. Farmer preference has shifted the emphasis away from community-funds to personal ones. Technical assistance has been provided by Project technicians.

RFs have been a useful tool for community agricultural committees in providing a focus for their activities. They have increased technology adoption, specifically in areas of Project emphasis. Limitations were encountered in administrative capacity of community leaders, and the need for external supervision.

Benefits from RFs perceived by farmers varied. Depending on the community, seed availability, animal upkeep, or income generation was cited as the main result.

RF viability has improved over time. Input recovery ranged from 50 to 90%. Low recovery was always related to climatic conditions, such as drought or frost. Flexible recovery terms were an important feature to recipients.

Income to aggregate investment in RFs by the Project was negative in three of the six years of operation. Total investment over this period was equivalent to US$298,000. Even with negative income in some years, the aggregate internal rate of return is estimated to be about 30%. This indicates that farmers are deriving benefits, though does not imply that the Project is generating extra income.

1. INTRODUCTION

This report documents experiences with seed revolving funds (SRF) in highland peasant communities of Peru. The objectives of the report are: 1) to describe briefly the characteristics, constraints and strategies used in organizing and implementing SRF, and 2) assess the reaction of the communities involved in the PISA Project to the SRF in the longer term. The study covers four cropping cycles from 1985 to 1989.

The PISA Project has experimented with different types of revolving funds in highland communities (including the purchasing and loaning of recommended agro-chemicals); however the report will be limited mainly to the SRF.

2. BACKGROUND

Agricultural credit in Peru comes from the Government (90%) and is channeled by the Development Bank (Banco de Fomento) through the Agrarian Bank (Banco Agrario). About 10% is provided by commercial banks. Other credit resources come from the informal sector like family and friends, input stores, and credit advances made by middle-men and retailers.

The last national survey of rural homes conducted in 1984 showed that only 8% of the total number of farmers in Peru received loans from the Agrarian bank and that the majority of loans went to farmers in the coastal and jungle regions growing cotton, rice, coffee and maize. Farmers who obtained credit tended to be better-off, had irrigated lands and more access to formal education.

One way to make credit available to poor highland communities has been the creation and operation of revolving funds. Revolving funds have been organized by NGO's and supported by international cooperation institutions who often provide the necessary initial investments. Loans may be in cash or inputs such as seeds, fertilizers, and pesticides. If credit is given as cash often the money is used by the farmers for other purposes.

Revolving funds commonly show resource depletion due to a) negative interest rates and b) high rates of inflation. Poor management could be another important factor. When interest rates are zero, like in the case of the PISA Project, the revolving funds are obviously subsidizing the farmers. If interest rates are lower than inflation a resource depletion is to be expected. In SRF the real rates of return are often better given that farmers pay back with seeds.

After the very serious droughts of 1982 and 1983 several experiments using revolving funds have been conducted in Puno, the goal is to make some capital available to farmers so they can cope with the emergency and immediate production needs. Most loans from revolving funds provided inputs to farmers and avoided use of these resources for other needs.

3. METHODOLOGY

This study considered the following steps:

3.1 A survey conducted with field technicians directly responsible for the operation of the community SRF

3.2 A survey conducted with farmers (recipients and non-recipients) to assess their views on advantages and limitations of the SRF

3.3 Economic and financial analyses including costs and benefits, internal rates of return, and net actual values.

4. IMPLEMENTATION

4.1 General aspects:

The revolving funds are intended as a community service to make available improved seeds, veterinary products, fertilizers and other inputs, and money in special circumstances. Seeds have been produced at Illpa Experimental Station and other stations of INIAA in order to assure good initial quality.

In the 85/86 cropping season the PISA Project started implementing the SRF in four communities, namely: Llallahua, Jiscuani, Kunurana Bajo and Viscachani and under two different modalities: community and personal loans. Personal loans also have two modalities: with and without technical assistance.

The seed nurseries are established in community lands or in areas loaned by private farmers. Community members carry out the land preparation, seeding, cultural practices, harvest and other activities as necessary. The Project provides improved seeds, fertilizers and other agro-chemicals as well as technical recommendations including seed rates, fertilizer use (both chemical and organic), minimum pesticides application, selection of seed categories at harvest, storage in diffuse light structures, etc.

After harvest the community pays back with selected seed of similar quality and cash, at prevailing market prices, for the inputs. The rest of the seed is used to plant new areas, cover hand labor costs and pay for other community needs. Each community member who participated and contributed to the implementation of the SRF gets a share, normally seed for new plantings or for home consumption.

Loans for personal seed nurseries may or may not include technical assistance. In both cases the Project provides the necessary inputs.

Animals play an important role in the communities, either as a main component of the peasants production system or as a key supplement to their activities. Animal production is constrained by internal and external parasites and therefore the PISA Project organized community revolving funds to purchase and maintain a minimum of veterinary inputs aimed at the control of parasites.

4.1 Beneficiaries of the SRF:

The SRF are obviously not enough to have a total coverage of the communities seed needs. The following problems were encountered:

- Some farmers, arguing that their planting areas were larger or that they were community leaders, insisted in getting more improved seeds than they could possibly manage.

- The poorer farmers were often shy on their requests.

- Sometimes husband and wife would make separate requests for seeds or the family head would apply for seeds under the two schemes: with and without technical assistance.

Considering the above difficulties the Project gave a higher priority to community SRF than to individual SRF and also to those recipients accepting technical assistance.

4.2 Planning and management of seed nurseries and inputs use:

The distribution of benefits (seeds, inputs) is done in meetings attended by the Presidents of the Community Council of Administration and the Agricultural Committee. A contract is drawn up and the two Presidents act as guarantors both to the community and to individual farmers.

The animal production revolving funds started in August 1985. Technical back-up on specific animal health problems has been provided by the Project's resident technicians.

5. FIELD TECHNICIANS SURVEY RESULTS

Field technicians implemented the SRF during four agricultural seasons. Starting with the 89/90 season more responsibilities have been assumed by the communities themselves. The survey included 34 questions, however, the results are summarized and discussed under four topics, as follows:

5.1 Types of revolving funds:

As highlighted before, the PISA Project has experimented with a number of revolving funds. Table 1 provides location, number and types of revolving funds and indicates the diversity and priority activities in each community. It is clear from the Table the SRF involving crops and animals are the most important ones.

Table 1. Number and type of revolving funds

Community	Number	Type
Anccaca	3	Crops, animals, handcrafts
Apopata	4	Crops, animals, marketing, health
Jiscuani	5	Crops, animals, carpentry, health, community store
Kunurana	2	Crops, animals
Llallahua	4	Crops, animals, community store, health
Luquina	3	Crops, animals, health
Sta. Maria	7	Crops, animals, community store, carpentry, metal work, handcrafts and health

The SRF emphasized cereals and grain legumes, this is shown in Table 2.

The relative importance of each crop changes depending on the ecology and interest of the community involved. However, except for Apopata where the emphasis is on animal production, potatoes and quinoa (*Chenopodium quinoa*) are key components of the production systems in the other communities. These two crops are part of the daily diet of community families and are given a high degree of importance by INIAA (National Agricultural Research Institute). The Project did not provide support to Andean tubers such as oca (Oxalis tuberosa) and olluco (Tropaeolum tuberosum) despite the interest of the farmers and requests from the Projects's technicians.

Due to pressures from the farmers themselves there was a gradual shift from community to individual SRF.

Table 2. Seed distribution by community

Community	Seed
Anccaca	Potato, quinoa
Apopata	Wheat, clover
Carata	Potato, quinoa, barley, faba beans, canihua, oats
Jiscuani	Potato, quinoa, canihua, oats, wheat, alfalfa barley, faba beans
Kunurana	Potato, quinoa, canihua, faba beans
Luquina	Potato, barley, wheat, faba beans
Sta. Maria	Potato, quinoa, barley

5.2 Organization and management:

The majority of the field technicians felt that the lack of clear rules and regulations constrained the implementation and management of the SRF (Table 3). For instance there was a tendency for the smaller amounts of seed loans to be considered as "gifts" by some community members.

Table 3. Organizational aspects of the SRF

Community	Regulations	Criteria to select beneficiaries	Coverage %
Anccaca	No	Not clearly spelled	80
Apopata	No	Not clearly spelled	45
Jiscuani	No	Not clearly spelled	30
Kunurana	No	Not clearly spelled	50
Llallahua	Yes	In coordination with Agricultural Committee	80
Luquina	No	In Coordination with Agricultural Committee	55
Sta. Maria	Yes	Started later in the process	85

5.3 Advantages and disadvantages of the SRF:

The perceived advantages of the SRF could be summarized as follows:

- SRF are interest free
- SRF are hassle free (administratively) and save time
- There are not strong penalties for the beneficiaries should they not be able to fulfill their obligations
- SRF include seeds and other inputs and therefore are used for the intended purposes
- SRF allow for other relevant technical assistance to farmers

Field technicians believe community members truly appreciate the above listed advantages and would be willing to make every effort to maintain and improve the SRF schemes.

The experience with the SRF indicate there are other important advantages, specifically they seem to help strengthen the communities' organization and tend to increase the adoption of technology. There are also some perceived limitations. A summary of benefits and limitations is shown in Table 4.

Table 4. Benefits and limitations of the SRF

	Benefits		
Community	Organization	Improves Technology Adoption	Limitations
Anccaca	Improves	Yes	Lacks administrative capacity
Apopata	Improves	Yes	Lacks administrative capacity
Carata	Improves	Yes	Lacks administrative capacity
Jiscuani	Improves	Yes	Lacks administrative capacity
Kunurana	Improves	Yes	Lacks administrative capacity
Llallahua	Improves	Yes	Needs stricter control
Luquina	Improves	Yes	Lacks adequate organization
Sta. Maria	Improves	Yes	Lacks supervision and penalties

The SRF have contributed to create or to re-activate the communities' agricultural committees by giving a new sense of purpose to them. On the technology adoption side there are indications of good acceptance of the new varieties introduced by the Project, particularly common and bitter potatoes, and quinoa.

The successful re-discovery and utilization of ancient high-ridge cultivation combined with the advantages of the SRF have brought added benefits to the Carata community. Other successful examples include the combination of SRF and low-cost diffuse light storage structures, which have been adopted by several communities.

According to field technicians the most serious limitations relate to the following:

- The relative low administrative capacity of the community leaders to handle the SRF
- The need for external supervision

Community members themselves seem to be inclined for a continuation of some "outside" management or supervision. The alternative, recognised by the Project, is to conduct more training activities in administration.

6. COMMUNITY MEMBERS SURVEY RESULTS

This survey included five of the eight communities involved in the PISA Project.

6.1 Technical assistance:

Community members have received advice and support on crops, vegetables and animal production. A high percentage of farm families have benefited (Tables 5 and 6).

Table 5. Technical assistance in the communities involved in the PISA Project

Communities	% of farmers
Anccaca	75
Apopata	93
Carata	85
Jiscuani	90
Santa Maria	88

Table 6. Categories of technical assistance in %

Categories	Anccaca	Apopata	Communities Carata % of farmers	Jiscuani	Sta. Maria
Crops	6	0	41	27	21
Animals	6	33	0	10	6
Crops-animals	56	47	35	43	38
Vegetables	0	0	0	0	3
Others	6	13	12	7	18
None	0	0	3	0	9
Does not know	25	7	9	13	6

The Project's experience indicate that farmers do not adopt rapidly or completely a new technology due to risk and economic factors. However given the long presence of the Project, over five years now, the rate of technology adoption is comparatively high (Table 7).

Table 7. Percentage of farmers who have followed technical recommendations

Communities	% of farmers
Anccaca	63
Apopata	93
Carata	93
Jiscuani	57
Santa Maria	93

Possibly one of the reasons for high adoption is the fact that the Project has located technicians directly in the communities. Those who have not adopted new technologies claim that the costs and few follow-up visits by the technicians are the main causes. Project staff have learned that farmers do not usually modify recommendations involving veterinary products since they are aware that lower or higher dosages may bring counter-productive effects. However farmers do tend to adapt and modify crop production techniques in order to suit them to their own circumstances.

6.2 Coverage and availability of seeds and other inputs:

Farmers not participating in the SRF have expressed fear of debts as the main cause for their reluctance. A breakdown of seed and input availability in each community is shown in Table 8.

Inputs available to a particular farmer do not cover his or her total needs, the range is between 19 to 54%. This is presented in Table 9 along with reasons and explanations. The timing for seed and inputs availability is critical. Over 90% of SRF beneficiaries indicated that timing was appropriate.

Table 8. Percentage of farmers who had access to seeds and other inputs

Inputs Community	Seed[1/]	Fertilizers	Pesticides	Veterinary Products
Anccaca	81	50	25	6
Apopata	10	0	0	100
Carata	100	14	0	0
Jiscuani	67	23	7	0
Santa Maria	96	68	14	0

[1/] Largely potatoes

Table 9. Reasons for input demands not being met

Communities Reasons	Anccaca	Apopata	Carata	Jiscuani	Sta.Maria
	Percentage of farmers				
Insufficient seed availability	20	0	11	13	23
Farmer had some land ready	10	0	33	75	31
Too many beneficiaries	60	0	11	12	0
Lack of financial resources	0	13	6	0	0
Inputs not enough for all	0	75	33	0	38
Unavailability of specific inputs	0	13	0	0	0
Others	10	0	6	0	7

6.3 Benefits, problems and recommendations:

The percentage of farmers quoting specific benefits from the SRF are presented in Table 10. The benefits, as expected, vary depending on the community. In Carata and Santa Maria the key element has been improved seeds, while in Apopata the SRF allowed the community to better upkeep their animals, in Anccaca and Jiscuani income generation was cited as an important result.

Table 10. Benefits obtained from SRF

Reasons	Communities Anccaca	Apopata	Carata	Jiscuani	Sta. Maria
	Percentage of farmers				
Increased income	54	0	0	55	14
Increased production	8	0	32	5	29
Higher yields	0	0	27	0	4
Access to improved seeds	0	0	18	0	18
Better animal upkeep and weight gain	0	80	0	0	0
Disease prevention	31	10	0	40	0
Availability of inputs	0	10	0	0	0
None (89-90 season)	0	0	18	0	18
Did not respond/know	8	0	5	0	18

During the 89/90 season drought and frost hit the Puno region and caused very severe crop losses, therefore for many farmers the use of improved seeds did not increase production and augmented their debts. In Apopata 10 % of beneficiaries felt that use of veterinary products did not bring about gains in animal production and animal health.

Farmers in different communities expressed their ideas on specific improvements that could be made in the SRF. These suggestions ranged from allowing use of traditional animal medicine in the SRF to encourage further participation of individual farmers. Their views are summarized in Table 11.

Table 11. Recommendations to improve management of SRF

Reasons	Anccaca	Apopata	Carata	Jiscuani	Sta.Maria
	Communities				
	Percentage of farmers				
Do not know, do not answer	0	30	0	10	4
Include other animal species in SRF	8	0	0	0	14
SRF's to continue as is	38	0	27	40	29
Increase availability of improved seeds	23	0	14	0	0
Increase availability of veterinary products	8	10	5	0	4
Expand SRF to private growers	0	0	23	0	0
Visits by technicians should be more frequent and include all farmers	0	10	0	0	4
Others	23	50	32	50	46

6.4 Recovery and viability of SRF in difficult years (drought):

The recovery of SRF has improved over the years, probably as a result of the learning process. Recovery has ranged from 50% to 90% and beneficiaries indicate two main reasons to comply a) they feel a moral and legal commitment and b) good production in "normal" years (Table 12). The causes for low recovery, as clearly expressed by farmers, are always related to climatic problems like drought and frost. For those harsh years farmers offer a number or remedies and suggestions including extension of payment deadlines and fresh loans (Table 13).

Table 12. Reasons given for recovery of SRF

Reasons	Anccaca	Apopata	Carata	Jiscuani	Sta.Maria
	Communities				
	Percentage of farmers				
Good production (good and normal years)	0	0	64	50	44
Must comply with commitment	100	89	29	40	46
Other	0	0	7	10	0
Does not know, does not answer	0	11	0	0	0

Table 13. Farmers' suggestions about SRF management on drought years

Suggestions	Anccaca	Apopata	Carata	Jiscuani	Sta.Maria
	Communities				
	Percentage of farmers				
More payment deadline/year	46	10	41	40	57
Scrap debt	0	0	18	5	0
More payment deadline and provide fresh loans	15	0	41	20	14
Provide vitamins and balanced diet for animals in SRF	0	70	0	0	0
Provide improved pasture seeds	15	10	0	5	0
Increase grazing areas and provide vitamins for animals	0	10	0	0	0
Other	23	0	0	20	21
Does not know, does not respond	0	0	0	10	7

Community members would be quite willing to continue using the SRF, because SRF bring benefits and are convenient during good and "normal" years. This is reflected in Table 14.

Table 14. Reasons given by farmers to continue using SRF

Reasons	Anccaca	Apopata	Carata	Jiscuani	Sta.Maria
	Communities				
	Percentage of growers				
Does not know, does not answer	0	10	0	0	0
SRF are only access to improved seeds	0	0	5	0	3
Brings benefits in good and normal years	100	60	86	93	86
Interest should be charged to crops inputs but not to veterinary products	0	10	0	0	0
Other	0	20	9	7	11

Farmers expressed the view that if loan payments were very strict (like those imposed by the State Banks) they would not participate in the SRF schemes. In such a case they may look elsewhere, particularly to informal credit sources. Surveys clearly revealed that farmers look for flexible, economic and timely credit systems.

6.5 Views of non-participating farmers:

The range of farmers who have had no access to SRF is 16 to 37%. The reasons include "having too small plots", "seed is not enough for all", "high cost of veterinary medicines", etc. These findings are shown in Table 15.

Table 15. Reasons for not participating in SRF

Reasons	Communities Anccaca	Apopata	Carata	Jiscuani	Sta.Maria
	Percentage of growers				
Community SRF are insufficient for private growers	19	0	58	20	0
Insufficient for all	44	0	25	56	0
Too small plots. Own seed is enough	0	0	8	0	67
Was absent when SRF were implemented	0	40	8	0	33
Inputs are expensive and prefer local remedies	0	40	0	0	0
There was little publicity about SRF	38	20	0	23	0

Many farmers expressed that even they did not participate the SRF are a positive development.

7. SRF AND TECHNOLOGICAL CHANGE

SRF can be seen in at least two different perspectives: as a means of credit and as a means of technological change.

7.1 Seed and fertilizer use and adoption of new varieties:

An example of plant density recommendations and those finally used by farmers is presented in Table 16. It appears that the adoption of the recommended practice was not very rigorous. The reasons given for the variation include the total availability of seed at planting time and the farmers' own experiences.

Table 16. Degree of acceptance of plant densities

Community	Recommended density	Plant density used per ha	Range
Jiscuani	1500	1515	1354 - 1758
Llallahua	1500	2356	2144 - 2625
Anccaca	1500	1033	806 - 1267
Sta Maria	1500	1471	1188 - 1640
Carata	1500	1326	1164 - 1410

There is a clear tendency in the communities to adopt new seeds, particularly potatoes and quinoa (Table 17). It is important to state that adoption of new varieties does not imply giving-up the traditional ones. What it means is that farmers, as always they have done, mix the new and the traditional germplasm. This strategy has increased over the last five years when PISA Project varieties became available. There are also differences on what is adopted, for instance bitter potatoes are more readily adopted than regular potatoes, perhaps due to inappropriate research focus in the past resulting in varieties without the required characteristics. Quinoa is a different story: appropriate varieties are now available even with less overall research efforts.

The degrees of adoption reached seem to point out the advantages of the SRF as compared to other extension strategies. Clear examples of high adoption levels have occurred with potato varieties Andina and Piñaza and quinoa varieties Kancolla and Blanca de Juli.

The results of fertilizer SRF are more variable as indicated in Table 18. Fertilizer recommendations are more difficult for farmers to follow, possibly due to the higher costs involved, the common unavailability of these inputs in Puno, and the SRF requirement that fertilizer loans be paid with cash thus increasing indebtness risks.

Table 17. Adoption of improved seeds

	Anccaca		Jiscuani		Ilave	Azangaro
	1985/86	1989/90	1985/86	1989/90	1989	1989
Potato	35	40	15	35	S.I.	N.A.
Quinoa	0	70	S.I.	30	90	N.A.
Cunihua	0	N.A.	*	*	N.A.	40
Barley	0	*	0	N.A.	N.A.	N.A.

* Crop is not important
N.A. Information not available

Table 18. Acceptance of fertilizer recommended rates

Community	Recommended rate	Rates actually used	N range used
Jiscuani	120 - 100 - 80	92 - 99 - 60	57 - 140
Llallahua	120 - 100 - 80	107 - 96 - 74	38 - 176
Anccaca	120 - 100 - 80	29 - 32 - 24	19 - 36
Sta Maria	120 - 100 - 80	67 - 80 - 70	58 - 79
Carata	120 - 100 - 80	N.A.	N.A.

7.2 Influence of improved seeds and fertilizers on productivity:

Two production functions have been developed in order to get an indication of the importance of improved seeds and fertilizers (Table 19). The plant density (D) and fertilizer (NPK), particularly P, account for a large percentage of yield variability.

Table 19. Plant densities and fertilization as production functions

Production function	Equation	R^2
Lineal	$R = 7034 + 1.1D - 141N + 338P - 0.2K$	0.81
Quadratic	$R = 11307 + 0.001D^2 - 0.6N^2 + 1.4P^2 - 1.18k^2$	0.79

7.3 Traditional technologies:

In the community of Carata there is a strong correlation between SRF and the present strategy of increasing the area planted to high ridge systems (Huaru Huarus). Farmers plant the improved seeds in the highly successful Huarus, SRF recovery is also high in this community.

8. LEVELS OF SRF RECOVERY

Seed recovery data is breakdown for cereals and tubers. Economic evaluations include internal rate of return (IRR) and net actual value (NAV).

8.1 Seed recovery by season and community:

Seed recovery has been increasing as shown in Table 20, variations are due mostly to climatic factors. The 1990 season has been characterized by severe drought, floods and frost and the data is not yet available.

Table 20. Recovery of cereal and tuber seeds

	1985/86	1986/87	1987/88	1988/89
Tubers	90	62	65	99
Cereals	N.A.	69	93	91

Seed recovery rates of the PISA Project are higher than those reported by the CEPIA Project in Puno from 1983 to 1986. The figures for the CEPIA Project ranged from 22 to 72%.

Tables 21 and 22 show the specific variations in each community. The differences are largely a reflection of the organizational capacity of each community and the degree of supervision and follow-up of the Project's technicians.

Table 21. Average recovery of cereals and tuber seeds from each community

Community	Tubers	Cereals
Jiscuani	67	83
Luquina	31	71
Llallahua	60	68
Kunurana	84	N.A.
Anccaca	58	100
Sta. Maria	89	88
Carata	100	N.A.
Uray Ayllu *	100	100
Puna Ayllu *	100	65
Vizcachani **	73	40

* One season
** Average of two seasons

Table 22. Percentage of seed recovery from SRF in ten communities and four cropping seasons

Community	1985/86	1986/87	1987/88	1988/89	Average by community excluding 1989/90
Jiscuani					
Tubers	100	22	67	80	67
Cereals	N.A.	N.A.	66	100	83
Luquina					
Tubers	N.A.	35	27	S.I.	31
Cereals	N.A.	42	100	N.A.	71
Llallahua					
Tubers	67	40	60	74	60
Cereals	16	N.A.	100	88	68
Kunurana					
Tubers	94	78	79	N.A.	84
Cereals	N.A.	N.A.	N.A.	N.A.	N.A.
Anccaca					
Tubers	N.A.	40	28	106	58
Cereals	N.A.	100	100	100	100
Santa Maria					
Tubers	N.A.	57	96	113	89
Cereals	N.A.	N.A.	100	75	88
Carata					
Tubers	N.A.	100	99	120	106
Cereals	N.A.	N.A.	N.A.	N.A.	N.A.
Urac Ayllu					
Tubers	N.A.	100	N.A.	N.A.	100
Cereals	N.A.	100	N.A.	N.A.	100
Puna Ayllu					
Tubers	N.A.	100	N.A.	N.A.	100
Cereals	N.A.	65	N.A.	N.A.	65
Vizcachani					
Tubers	100	46	N.A.	N.A.	73
Cereals	N.A.	40	N.A.	N.A.	40
Promedio/campana					
Tubers	90	62	65	99	
Cereals	N.A.	69	93	91	

8.2 Investment structure and coverage:

The Project's investment in SRF (crops) were 61, 74, 89, and 58 % for the 85-86, 86-87, 87-88 and 88-89 seasons respectively. This is indicated in Table 23.

Table 23. Investments by activities (in Intis)

Investments	1985/1986	1986/1987	1987/1988	1988/1989
Crops	43,509	405,264	515,659	3,258,190
Animals	9,373	133,574	62,826	2,331,004
Health	0	6,493	0	0
Total	52,882	545,331	578,485	5,589,194
% invested in crops	61	74.2	89	58

By far the largest investments were in potato seed production (Table 24). Potato is the most important crop and normally a daily staple. Investments in animal and human health have been marginal. The Project has worked with up to twelve communities in different seasons, however, over the last two years there has been a concentration of efforts in only five communities.

Table 24. Characteristics of SRF

	1985/86	1986/87	1987/88	1988/89
Kg of potato seed in SRF	9900	54,295	37,440	32,589
Area (ha) of potatoes	6.80	45.83	24.98	21.73
Area (ha) other crops	3.38	49.13	17.11	14.19
% of area under potatoes	33	52	41	40

8.3 Estimates of costs and income:

The SRF include costs associated with the Project itself and those associated with the farmers. Besides direct costs of inputs, the Project considers administrative, technical assistance, land rental and financial costs as well. Most of the income is derived from potatoes, followed by other crops and animal production activities.

Income was negative during the first two years and also in the 89/90 season. It showed a positive balance in 1986, 1987 and 1988, as highlighted in Table 25.

Table 25. Costs and income of the SRF (in USD)

	Agricultural seasons					
	1985	1986	1987	1988	1989	1990
COSTS						
1. Inputs	3,456	30,810	12,308	13,134	7,258	7,258
2. Technical assistance	316	3,406	1,718	1,199	663	663
3. Administration	1,013	6,243	4,824	3,848	2,126	2,126
4. Farmer's hand labour (potato)	4,332	17,216	24,797	16,463	9,098	9,098
5. Oxen (for potatoes)	1,558	9,605	7,421	5,920	3,271	3,271
6. Hand labor & oxen costs for other crops	2,010	17,219	12,597	7,637	4,221	4,221
7. Land rental	935	5,763	4,453	3,552	1,963	1,963
8. Financial costs		817	5,416	4,087	3,105	1,716
TOTAL	14,436	95,677	72,204	54,858	30,316	30,316
9. Total income from potatoes		12,658	135,301	46,954	47,827	2,870
10. Total income from other crops		1,392	23,443	6,780	6,803	408
11. Income from animal production		1,102	13,583	2,406	15,953	1,595
TOTAL INCOME		15,152	172,327	56,142	70,584	4,873

Sources: Prepared by authors based on:
1. Primary information from the PISA Project files
2. Accounting books of the Project
3. Files and field books

8.4 Economic analysis:

Costs and benefits are indicated in Table 26 for a nine year period. Data for the last three years is an estimate of the authors.

Table 26. Internal rate of return, actual net value and costs/income values

YEAR	1985	1986	1987	1988	1989	1990	1991	1992	1993
Total income		15,152	172,327	56,142	70,584	4,873			
Total costs	14,436	95,677	72,204	54,858	30,316	30,316			
Net income	-14,436	-80,525	100,112	1,265	40,268	25,443	20,000	20,000	20,000
Internal rate of eturn		0.30							
Actual net value		49,637							

The internal rate of return for the Project's SRF is estimated as 30% when calculated in US dollars. It is a very good return given that the actual interest paid by commercial banks has been around 6%. The above data indicates that farmers are deriving benefits from the SRF, however the data does not imply that the PISA Project is generating extra income given that the seed recovery rate is not 100%, as previously discussed.

The actual net value, the difference between costs and income in actual terms, showed a positive balance for the Project of $49,637 US. Therefore it may be concluded that the SRF have achieved overall positive returns, despite the serious climatic drawbacks of the 89/90 agricultural season.

9. DISCUSSION AND CONCLUSIONS

Seed Revolving Funds are useful seed dissemination mechanisms that contribute to increased productivity by providing key inputs and subsidizing very poor farmers with low interest rates.

The SRF are also an intermediate credit scheme with a number of advantages over both the official and the informal systems, one of those advantages is loaning of inputs instead of cash in order to diminish risk of funds being diverted to other activities.

Community members have expressed satisfaction about SRF being an effective credit alternative capable of having an impressive coverage of 40 to 80% of the total community members. They have also indicated that SRF allowed them to gain timely access to improved varieties and chemical inputs.

Indirect benefits of SRF include enhanced community organization and increased awareness of useful technological changes. Up to 90% of participating farmers followed, often with modifications to suit their needs and conditions, the technical recommendations. A case in point is the community of Santa Maria where technical assistance reached 88% of growers, improved seeds and inputs were made available to 96% and 68% of growers respectively and 93% of the participating farmers followed technical recommendations.

The SRF contributed to improve potato yields by 67% in the participant communities. The comparison is based on Provincial averages from 1986 to 1989. The SRF have helped to increase improved seed multiplication and use at the community level. A total of 734 t of seed was grown in 8 communities, part of the seed produced has reached other communities outside the Project's domain.

The most serious limitations of the SRF are the following:

- Often farmers regard the seeds and inputs as a gift (given the relatively small amounts provided)
- The overall capability to administer the SRF is rather poor at the community level
- During years of extreme climatic variations, drought and frost in particular, the SRF can be severely depleted
- Field supervision by technicians has not been quite appropriate

There is a strong perception by Project staff that more education, management training, and better control are required to overcome the above listed problems.

Confounding factors affecting appropriate controls included the fact that some of the seeds grown in the communities had to be taken to Puno for storage, given the poor existing infrastructure in some communities. More recently diffuse-light storage rooms have been completed in all participating communities.

There is a clear positive tendency on loan recovery of the SRF. The rate of recovery for potatoes and bitter potatoes went from 62% in the 86/87 season to 99% in 88/89 (the average is 80%). Likewise for cereals the figures were 69% to 91% for the same periods. These are much higher than the reported figures of the CEPIA Project: 57% on the average for potatoes during the years 1983-1986. The economic performance of the SRF has also been positive. The Internal Rate of Return has been determined as 30%, a high value indeed.

Perceptions of the SRF varied and often presented distorted pictures. In the short-term farmers seem to be clear about the benefits, however in the long-term some farmers felt that others, e.g. the project itself or its representatives, could be the direct beneficiaries. It is extremely important to discuss before hand with community leaders about which revolving funds should be implemented. Only those with good potential and strong endorsement and commitment should be pursued. It must also be understood that the SRF are supplementary and not a substitute to other types of credit.

Most beneficiaries feel that evaluations of SRF must be done for several seasons to account for climate and other variations. This is illustrated by the severe droughts of 1989-1990 that devastated the Puno highlands. The PISA Project has decided to continue supporting the SRF to somewhat counteract the ravages of climate in the area.

THE ROTATIONAL SEED FUND : A CASE STUDY OF SEED POTATO PRODUCTION AND DISTRIBUTION IN PUNO, PERU

Jorge Reinoso
Roberto Valdivia

PISA Project,
National Institute for Agricultural and Agroindustrial Research
Puno, Peru.

Abstract

A seed revolving fund (SRF) has been developed as a component of the Seed Production Program, Illpa Zonal Experiment Station, Puno, Peru. The principal crop addressed is potato, which is indigenous to this region.

The majority of the Department of Puno lies at an altitude above 3,800 masl. Climatic conditions can be extreme, and only close to Lake Titicaca, with its thermoregulatory effect, are small-farm production systems well-developed and relatively free from constraints.

Of about 120,000ha under cultivation, about 40,000ha is dedicated to potato. Half of this would be in production units of less than 5ha. Of the area seeded to non-bitter types, 15% is the variety Andina, 83% are native varieties, and 2% are other introductions.

The SRF was conceived of as a mechanism to stimulate seed multiplication and dissemination by the private sector. Difficulties in monitoring these processes led to a retraction of the SRF, so that it now operates within the Illpa Station as the main institutional seed production activity. Initially intended to be self-financing, the SRF is still dependent on an annual injection of donor funds.

INIAA, the national-level agency of which Illpa is a part, viewed this SRF as a model for similar funds for other commodities, anticipating returns that would at least fund part of the research program in each case. Where established, these other funds have experienced similar difficulties.

The potato SRF has distributed a total of 1,346t of seed in the period 1986-90, via various channels. Of these, direct distribution by Illpa Station has been the most significant, followed by externally-funded projects and NGOs. The latter have been more successful in promoting concomitant technological change.

The economic crisis in Peru has been a major factor in the limited success of the SRF. Annual losses to inflation have been between 19-34% of SRF income. Initially, institutional regulations barred the SRF from taking countermeasures. Similarly, public-sector management of the SRF was not sufficiently attuned to timely payment by recipients of seed, resulting in continual cash-flow constraints.

Successful operation of the SRF in future requires not only attention to financial management, but also determination of the actual benefits that accrue to the fund. Equally, the SRF must contribute to solving small-farm production problems, and be responsive to feedback from such sources. Studies of varietal diffusion and contribution to production through the SRF are needed.

INTRODUCTION

Different studies of agriculture on the Peruvian Altiplano have identified seed availability of the different crops as being a critical constraint to production increases. This does not only relate to factors such as variable climatic conditions, types of producers, commercialization, and to the genetic material of the main commercial crops of this region, but also to promotion and development policies behind seed production.

Farmers in Puno traditionally obtain their seed from their own stores, from farm fairs, or by exchange within the community. Few farmers can obtain improved seed, as this is beyond their limited means.

On the Altiplano, we have been trying to develop and institutionalize improved seed production by means of a rotational seed fund (RSF). This is an attempt to support the development of technology which is transferable to small farmers through extension programs, using the multiplication of germplasm as the vehicle. Since 1985, INIAA (the National Institute for Agricultural and Agroindustrial Research), with IDRC support to the Andean Farming Systems Project (PISA), has conducted RSFs at the institutional and community levels. The latter aspect is reported in another paper presented to this workshop.

The present paper discusses the institutional RSF developed to strengthen the Seed Production Program, as a component of the Research and Extension Program, at the Illpa Zonal Experiment Station, Puno. The RSF covers seed production in potato (sweet and bitter varieties), cereals (winter and spring), legumes (such as broad beans and lupins), and native crops such as quinoa, cañihua, oca and olluco.

The present analysis is carried out after four years of existence of the RSF, and considers its successes and constraints, and the possibilities of its expansion to a regional level, and whether it is an approach usable in other parts of the country.

It is worth noting that Peru has been passing through very difficult times during the period under consideration, and that the achievements of the RSF should be viewed in the light of the administrative and financial constraints of the managing institution. The results achieved reflect the dedication of the researchers involved.

This report covers the potato only. This crop has been by far the largest in terms of the attention paid to it through the RSF, given that it is the major subsistence crop of the region. It is of prime importance in economic and nutritional terms. To include the other crops would have required more space and time than is available to us here.

ORIGIN OF GENETIC MATERIAL

The collection of the principal native varieties of potato in the Department of Puno was begun in 1973. At this time the evaluation of this material began, focusing on production under conditions of hail, drought and frost, and to a lesser extent, pests and diseases, all factors contributing to output in this region. Ten varieties were first considered at this time.

Since this time, approximately 17 varieties from other parts of the country have also been introduced to the region. The origin of this material was the collections and crosses from the Research Stations of Cusco and Huancayo, material which had been evaluated under the same criteria as that of Puno, though with perhaps more emphasis on pests and diseases, and less on climatic factors. In Puno, where growing conditions are severe, production stability across sites and years is a prime concern.

Table 1 shows the local and introduced varieties which have received most attention during the last 15 years. Much other material, e.g. from ICA, the Colombian Institute for Agriculture, and from Peru's own National Potato Program, has also been evaluated, but found wanting under the severe climatic limitations of the Altiplano.

Table 1. Experimental yields (t/ha) of local and introduced varieties of potato. Puno 1974-86

Variety	Source	Yield (t/ha)									
		74	75	76	78	79	80	81	82	84	85
Mariva	Huancayo	2.3	20.0	15.2	23.6	2.9	15.8	8.9	6.1	11.1	32.0
Mantaro	Huancayo	4.8	20.0	17.6	34.9		9.0				
TTConde	Huancayo	6.3	27.6	12.8	33.9	1.2		10.8	5.9	16.7	28.0
H. Cusco	Cusco	5.8		11.7	21.8						
Yungay	Huancayo	8.6	14.7	34.1							
I. Negra	Puno	6.8	19.8	12.6	26.1	4.2	14.9	11.1	3.4	15.0	27.4
Ccompis	Puno	4.6		11.1	30.3	1.9	18.4	12.0	3.1	14.5	25.5
I. Blanca	Puno	3.3			27.0	4.4	10.1	16.0	5.4	13.6	23.1
Sipeña	Cusco	3.5	23.6	10.1	30.0						
Andina*	Puno		26.9	18.0	36.7	10.3	26.8	20.6	5.8	16.8	27.6
Revoluc.	Huancayo			21.3	16.4	1.8	12.9	10.5	3.8		
Antarqui	Huancayo			15.2	29.2	4.9	13.8	16.4	8.0	16.4	46.9
M. Bast.	Cusco					3.8	11.0	15.0	4.5		
S. Imilla	Cusco	22.1					11.2	13.3	6.3		
Alcatarma	Huancayo		19.3								
Particip.	Cusco									17.2	30.8
Ollanta	Cusco					17.5	25.1				
Huaycha	Cusco									16.4	39.7
Huancayo	Huancayo									15.3	32.0
Tahuaqu.*	Puno									12.8	39.3
Sillust.*	Puno										29.1
B. Casas	Cusco										17.1

* Clones introduced and selected in Puno, now local varieties.
Source: Various research reports, INIAA-PUNO

Clear from this table is the tremendous variability between years in yields. This has led to a general grouping of annual production into three ranges, poor, average and good. Because of this variability, and the necessity of being sure that a variety will weather the worst season, it is necessary for any new accession to be tested for a minimum of six seasons.

Table 2 shows the mean yield of these varieties for the three main seasonal groupings. Also shown are the regression coefficients and R^2 values for yield across years of some of these varieties. The regression coefficient is used as an index of stability (IS), a value of less than 1 indicating more stability than a value greater than 1. The range shown is from 0.65 to 1.55.

Table 2. Average yields of some varieties during poor, average and good production years.

Variety	Mean yield (t/ha) by type of year Poor year	Average year	Good year	IS	R^2
Mariva	3.7	12.7	25.2	1.05	0.99
Mantaro	4.8	13.3	27.5	1.11	0.99
TTCondema	4.5	13.4	29.8	1.24	0.98
H. Cuzco	5.8	11.7	21.8	0.78	0.99
Yungay	8.6	14.7	34.1	1.25	0.93
I. Negra	4.8	13.4	24.4	0.96	0.99
Ccompis	3.2	14.0	27.9	1.21	0.99
I. Blanca	4.4	13.2	25.0	1.01	0.99
Sipeña	3.5	30.0	23.6	0.95	0.50
Andina	8.0	20.6	30.4	1.09	0.99
Revolución	2.8	14.9	16.4	0.65	0.81
Antarqui	6.4	15.4	38.1	1.56	0.96
H. Bastidas	4.2	13.2			
S. Imilla	6.3	12.3	22.1	0.77	0.99
Alcatarma			19.3		
Participación		17.2	30.8		
Ollanta		17.5	25.1		
Huaycha		16.4	39.7		
Huancayo		15.3	32.0		
Tahuaqueña		12.8	39.3		
Sillustani			29.1		
B. Casas			17.1		
Mean	5.1	15.4	27.6		
S.E.	1.7	4.1	6.5		

Poor years: 74-75, 79-80, 82-83
Average years: 76-77, 80-81, 81-82, 84-85
Good years: 75-76, 78-79, 85-86

GENETIC BASIS

Native varieties in Puno are derived from sub-species *tuberosum* and *andigena* of *Solanum tuberosum*. Most of the introduced varieties also have this genetic origin, e.g. Andina is *tuberosum* x *tuberosum*, Tahuaqueña is *andigena* x *tuberosum*, and Sillustani is from *tuberosum* x *andigena*.

RESEARCH PRIORITIES

National research priorities have focused on increasing unit area yields, whether in the Coastal or Andean regions. This has been the case in Puno, where consideration of the extreme ecological conditions might have suggested a different focus. The reason for this common approach is that priorities were set at the national, not regional, levels. Improvement through breeding was seen as the most important activity, and within this, the selection of clones which express high yield potential when provided with high levels of inputs.

Considering the ecological restrictions on production in Puno, such as infrequent precipitation, low temperatures in critical periods, and hailstorms, research priorities should focus on varieties more stable over time, that is, it is more important for a farmer to have varieties with stable yields, than one with high yields in good years, and low ones in bad ones.

From this perspective, native varieties such as Imilla Negra and Imilla Blanca provide more promising genetic material than introduced varieties, in spite of the latters' yield potential. Still, there is a very strong tendency to focus on high yields, a direction exacerbated by use of significant fertilizer inputs in trials. Such use by small producers is limited, due to credit restrictions. Similarly, seeding date recommendations tend to focus on an optimum, rather than on responses to critical conditions beyond a narrow range of dates.

Another limitation has been accounting for qualitative factors. Native varieties present cooking qualities superior to improved varieties. As the majority of production is destined for home consumption, such quality is important to the producer.

One of the more recent advances in research has been the use of varietal mixtures. This is a common local practice. Such mixtures provide for more stable overall yields in such a marginal environment.

3. POTATO PRODUCTION

3.1 CHARACTERISTICS OF POTATO PRODUCTION UNITS

The Altiplano of Puno has four distinct agroecological zones (Table 3). Zone A, on the periphery of Lake Titicaca, has the best conditions for agriculture due to the thermoregulatory nature of the Lake. Crop and animal production is intensive, and potatoes and cereals both produce well. Zone B, further from the Lake, is the centre of concentration of bitter potato production, with minor emphasis on non-bitter potato and grains, and other Andean crops. Livestock production is more important in zone B than in A. Zone C is essentially livestock production, with some cultivation of bitter potato, kañihua, and native non-bitter potatoes. In Zone D, crop production is very limited, mainly to cereals for forage; livestock production is the main economic activity.

Table 3. Agroecological zones of the Altiplano

Zone	Annual precip. (mm)	Average annual temperature (C)	Altitude m
A (Lakeside)	800	10.2	3820-4100
B (Azangaro)	600	8.2	3950-4150
C (Altiplano)	700	7.6	3780-4200
D (Cordillera)	590	6.8	4200

Source: PISA (1988). Informe resumido.

Three types of production enterprise exist in these areas:

1. Enterprises resulting from the agrarian reform process, such as the SAIS (Sociedad Agricola de Interés Social) and CAP (Cooperativa Agraria de Producción).

2. Small and medium individual producers.

3. Peasant community small holdings.

In terms of land tenure, the first is found mainly in zones B, C and D, and occupies most of the cultivated area. Livestock is the principal production focus, with the exception of those units closer to the Lake. The small and medium producers are found in zones A, B and C. Peasant communities are found in all four zones, though the main concentration is in zones A and B.

The total area dedicated to cultivation in Puno varies by year between 110,000 and 140,000ha. Of this total, some 40,000ha is dedicated to potato production, 30,000ha to non-bitter and 10,000ha to bitter types. Almost half of the potato area is in units of 5ha or less.

Of the area seeded to non-bitter types, 15% is dedicated to the variety Andina, 83% to native varieties, and 2% to other introductions.

The nature of the production units strongly influences production technology. The size of land holding is a principal indicator of the social and economic condition of the producers. For example, peasant farmers with less than 1ha may have this spread among more than 10 parcels at different locations within the community. The increasingly larger individual producers and cooperative entities typically have fewer parcels.

Characteristics of parcel size, type of holding, and agro-ecological location together result in specific patterns of technology adoption. Improved varieties, with their dependence on chemical inputs and more sophisticated cultural requirements, find more acceptance among the larger cooperatives and individual producers, which both have access to credit and markets. The peasant communities rely on the native varieties, and the diffusion of improved material is more feasible here, though credit and marketing restrictions still constrain significant spread of such material.

THE ROTATIONAL SEED FUND

The success of an RSF depends partially on the ability of the originating breeders or station to produce and multiply sufficient material for subsequent diffusion. It should be stated that, when the RSF was originally conceived, it was intended that the majority of seed multiplication would be conducted by the private sector, and that this sector would be seen as the main agent in spontaneous adoption of improved varieties by a wider target group. Considerable difficulty was experienced in monitoring such production, and gathering the quantitative data necessary to evaluate the program. As a result, the RSF is now viewed as an institutional activity, with the resources of the Illpa Zonal Station (including its sub-stations) being used for seed production. The RSF has been supported financially by funds provided through IDRC from the Canadian International Development Agency.

Since its inception, the RSF has passed through different periods, which resulted in organizational changes. In this, it has mirrored the parent institution. There have been two main phases:

1. When the parent institution (INIPA, represented regionally by CIPA) undertook activities of both research and extension.

   This period can be sub-divided into the first year and subsequent years. In the first year, the RSF was run by CIPA from the Puno Experimental Station. The financial resources were provided by and administered by the PISA Project. In the second and third years, the RSF was transferred completely to CIPA.

   During this first phase, RSF activities included planning, monitoring, supervision and evaluation. The RSF was led by a Commission (subsequently a Committee), of Director-level staff of both CIPA and the PISA Project. This staff had the responsibility for programming, assignation of resources, management of the technology transfer and extension mechanisms, and price setting.

   During this period it would have been feasible to evaluate seed production at the farm-level, but for a variety of reasons it has not been possible to document the process.

2. When the parent institution (now INIAA) transferred extension responsibilities to the Regional Office of the Ministry of Agriculture.

   As a result of this transfer, from the fourth year, the RSF became distanced from technology transfer, resulting in significant difficulties in monitoring seed production and producer management. Links established with other institutions, such as NGOs, did not substitute for the original direct contact with the producers.

   In this fourth year, the RSF began increased support directly to the research program (genetic work, propagation of virus-free material, and agronomic studies), with some support for core activities of the Station. Because of severe institutional economic constraints, the RSF became an even more important source of operational budget for these program activities in the fifth year.

In the fifth year, in order to put a more formal stamp on the RSF, a separate management position was established, with its own administrative unit. Due to the difficulty of recuperating operating expenses through its own actions, the RSF still depended on an annual assignation of funds from the PISA Project.

The parent institution has taken the RSF experience, and established a national-level RSF program, designed to establish several regional RSFs in other parts of the country, which would define policy and objectives related to the production and supply of seed and inputs, pricing, and technology transfer. For reasons similar to those in Puno, these regional RSFs face the same problems in evaluating material disseminated at the field-level.

MATERIAL PRODUCED AND DISTRIBUTED BY THE RSF

Table 4 indicates the volume of material produced at the Illpa Zonal Station since the inception of the RSF.

Table 4. Area seeded and registered seed production achieved in the main varieties of potato produced by the RSF. Illpa Zonal Experiment Station 1985-90.

Year	85/86	86/87	87/88	88/89	89/90	Ave
Area cultivated (ha)	32	84	55	28	31	46
Total production (t)	605	390	449	216	239	379
Production (t/ha)	19	5	8	8	8	8
Production by variety (t/ha)						
Andina	23	5.6	9.9	14.9	7.0	10.6
Imilla negra	28	3.9	8.4	7.6	4.0	6.7
Ccompis	24	3.7	9.6	12.5	4.3	8.8
Imilla blanca	18	2.4	7.0	7.8	5.0	6.6
Tahuaque:a	37	9.7	14.8	2.2	2.7	13.2
Yungay		5.4	9.6	28.0	5.0	14.2
Sillustani		2.7	8.0	19.7	12.1	13.3
Huaycha				22.0	13.0	17.5
Valicha				20.0	4.0	12.0
Ave by variety	26	4.8	8.7	16.1	5.7	11.3

Source: RSF Reports, PISA-INIAA

It can be noted from Table 4 that both area seeded and yields obtained have been quite variable between years. Varietal differences are also evident.

Since inception, seed has been disseminated through the RSF by local Extension Agencies (1986-88), by the National Seed Service (SENASE, 1986), by various Projects and NGOs (1987-90), by producers (1986, 87 and 89), and by the PISA Project. Table 5 indicates the amounts of seed distributed.

Table 5. Seed distributed (t) by season and source through the RSF

Source	1986	1987	1988	1989	1990	Total
Illpa Station	165	95	81	51	101	493
Extension Agents	19	36	79	-	-	134
SENASE	26	-	-	-	-	26
Projects/NGOs	-	30	18	150	158	356
Producers	201	24	-	20	-	245
PISA	30	13	32	2	15	92
Total	441	199	210	222	274	1346

Source: RSF

In the first three years, the use of registered seed and improved management technology had little impact. This was due to:

1. The extension services and methods not being appropriate for the conditions of the target producers.

2. The genetic material distributed through the RSF being intended for seed multipliers, and not commercial producers. Unfortunately, it was distributed mainly among the latter.

However, not all channels showed the same results. The Projects and NGOs, and PISA, promoted the use of improved seed, using not only different extension methods, but also informal credit systems (mainly in-kind). These efforts have slowly brought a process of technological change, with there now being a preferential demand for seed from the RSF.

The experience of the last two years has shown an increase in this demand, especially from agencies which work in the area of extension and technology transfer.

The RSF has contemplated the extent to which it would be possible to meet the regional demand for seed potato. This would require collaboration between the RSF and local seed producers due to the volume of seed that has to be produced in order to meet regional needs. For instance, to meet 20% of the annual regional need would require 385ha of breeders seed from the RSF, and 1,500ha of certified seed from local producers. It is not infeasible to calculate that, through the RSF, it would be possible to change the variety grown throughout the region in only three years.

Shortly after its inception, the RSF was transferred from the PISA Project to the local counterpart agency (then CIPA, now INIAA), with the purpose of institutionalizing the fund, and testing the administrative systems created for its management. The main lesson that has been learned as a result is that the organization and administrative structure of state enterprises do not foster the entrepreneurial environment, nor the longer-term perspective necessary. State entities require that financial resources be managed through state banks, without the accrual of interest.

The hyperinflationary situation in Peru has also worked against the RSF. Losses due to inflation over the life of the RSF have averaged annually between 19-34% of income to the RSF. In an attempt to minimize these losses, it was decided to acquire crop inputs well ahead of time, to safeguard against continually rising prices. It was also decided to open a savings account for RSF purposes, which allowed interest to be accrued. Subsequently, a dollar account was established. However, these actions took several years to effect. Had these measures not been taken, annual losses could have reached almost 100%.

In state enterprises, money has only nominal value. It is of little importance to public servants that seed is not paid for on receipt, and that payment is arranged over time, or through intra-institutional loans, both of which, when settled in local currency, result in significant losses. In 1989/90, had seed been promptly paid for, the RSF would have been significantly better off, the resulting shortfall from late payment representing 54% of expected cash receipts during the year. This can be considered a direct loss in return to time and resources invested in research.

Over the life of the RSF, more importance has been given to cash flow than to an internal rate of return. Thus, the RSF has operated more as another program of the Station, where annual budgets are assigned for operation, and where the funds are expected to be spent. A positive aspect is that the RSF always had funds available for research or production activities, providing an essential resource to researchers with under-utilized infrastructure and no operating budget. It is estimated that the RSF provides up to 20% of the current budget of the Station in these areas.

While some attention has been given to administrative and financial management, it is important that more emphasis be put on analysis of the economic situation, so that possible losses to inflation could be achieved through different accounts in local and foreign currency in private banks. It is important to document in the current case that the danger to the fund comes not from basic technical or financial management, but from the monetary conditions in which the RSF is forced to operate.

COMMERCIALIZATION OF SEED POTATO

At the regional level, purchase of inputs and sale of farm produce is generally effected through local markets and businesses, both of which operate on the basis of the annual production cycle.

In the case of seed potato, and potato destined for consumption, the marketing cycles vary, but would respond directly to whether annual production is above or below average. Figure 1 shows the mechanisms that operate under normal conditions. When production conditions are favourable, regional production is capable of meeting regional seed needs. In such a case, seed is channeled principally through `rescatistas', or small-volume buyers, who sell on to larger wholesalers. The latter also sell on to regional seed houses. These intermediaries also acquire seed from other regions, such as Cusco and Huancayo, depending on the condition of those markets from year to year.

Figure 1 Seed Potato Commercialization Under Normal Circumstances

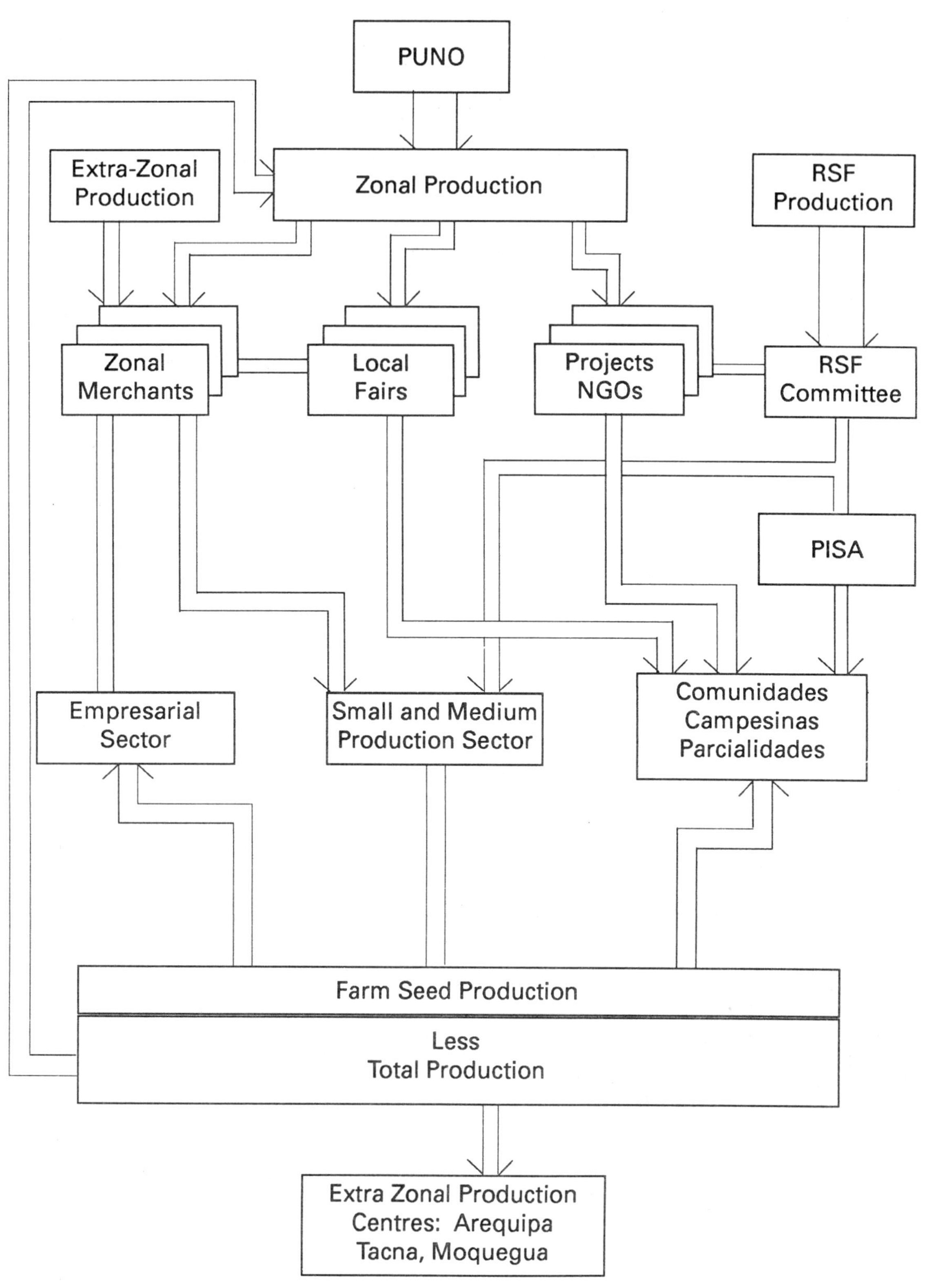

The regional houses will sell seed directly to the larger agricultural businesses (CAPs and SAISs), and to individual producers, though will also sell smaller volumes through local fairs in Juliaca and Ilave. Peasant farmers from campesino communities generally obtain their seed at these and other fairs, either through cash transactions, or by bartering.

The Department of Puno has received considerable support in recent years from a variety of development Projects and NGOs. These organizations also act as seed distributors principally to peasant communities, where in-kind credit mechanisms have been established. A significant proportion of the RSF material utilizes these channels. The RSF attempts to direct seed to agencies which have the capacity to monitor and evaluate the genetic material distributed.

In all these channels, seed quality is generally defined by variety and the size of the tuber. In some cases, seed origin may influence market price.

In poor production years, extra-zonal seed becomes of greater significance. As a result, the number of varieties available may expand, including both native and improved types. Between good and bad years, market prices may fluctuate by as much as 30%.

The potato being the principal crop in Puno, the question may be asked: Is it feasible to make a profit from commercial seed production? Given the existence of a strong market, the experience of the RSF has been that returns to production can vary between 30-50%. Given the greater empresarial nature of private enterprise, it is considered that commercial seed producers should be able to realize greater profits. Puno has traditionally been a source of seed for other Departments. With input from the RSF, it is possible that improved technology of production could lead to greater returns, e.g. with virus-free seed, yields could increase by 30%. However, the continuing tendency of local producers to stay with traditional varieties suggests that any future role the RSF may have in the regional seed industry should be based on increased output of these varieties, combined with a stronger technical assistance and credit program for commercial producers.

SUPPORT TO RESEARCH

The results of RSF activities are important to the research program. Producer yields indicate potential maxima in the use of improved technology, an importannt point of comparison with yields obtained experimentally. In this sense, the RSF is itself a research process. Obviously, both technical and economic analyses are important. Through such analyses, key variables affecting production may be identified.

SEED LEGISLATION

Seed legislation is defined in the General Seed Law - Law Decree No 23056, May 1980. Seed potato production is also governed by a specific Supreme Decree (105-82-AG and 036-83-AG). This legislation is intended to promote, standardize and control various activities, such as research, production, processing, and commercialization of seed throughout the country.

The Special Rules concerning Seed Potato are much more specific. These also attempt to define seed categories, a cultivar registry, seed certification and production, some aspects of comercialization, control and sanctions, and tariffs for the certification and labelling process.

While this legislation is intended to stimulate seed production, in reality it tends to impose a control on the process. No incentives are provided to those individuals or institutions dedicated to seed research, production or commercialization. Rather, the law hinders these processes through excessive bureaucracy, and sanctions for those who evade it.

MEDIUM-TERM PERSPECTIVES

In order to strengthen the RSF over the longer term, the following actions are necessary:

1. Restructuring of the financial management, given that the inflationary perspective in Peru does not permit the RSF to protect its capital with the given system of management.

2. Determine actual benefits, in order to show that the RSF justifies its existence.

3. Demonstrate that the RSF is an important element in the production of high quality seed material, e.g. virus-free seed, for use in the region.

4. Demonstrate that it is possible to solve the production problems of the small producer.

5. Assign or redirect resources directed to research, towards field-level evaluations as feedback to the RSF.

6. Conduct adoption studies at the farm level, to determine varietal diffusion and contribution to production.

7. Expand the training function of the RSF, not only in the area of agronomy, but also towards production systems, and their suitability in different areas.

CONCLUSIONS

In general terms, the most difficult aspect of the RSF has been that of inflation. The rules governing state enterprises prohibited in the early years the taking of decisions that would have allowed a measure of control over this problem. When such conditions exist, it is important for an RSF to buy its inputs well ahead of time.

It is also important to review the viability of an RSF under such marginal agro-ecological conditions. An RSF must be able to compensate for the extreme variability in production between years. Flexibility in area available for seeding is important, as is an adequate basic seed supply.

The experience in Puno has also shown that, to a degree, an RSF can not only finance seed multiplication, but can also generate resources for the research program, an important aspect when other sources of operating expenses are limited.

SWEET POTATO SEED SYSTEMS IN THE PHILIPPINES

Julieta R. Roa

Philippine Post Crop Research and Training Centre
Visayas State College of Agriculture

ABSTRACT

Sweet potato R and D received impetus with the creation of a nationally-mandated root crops center, thus recognizing the vast potential of the crop for food, feed and other industrial uses. Such rationale is strengthened as subsistence and marginal farmers are greatly affected; the improvement of their lot is a matter of national concern.

The initial major thrust centered on varietal improvement which spurred breeding activities first at the College of Agriculture and Institute of Plant Breeding of the University of the Philippines, College, Los Banos and more intensively at the rootcrops center at the Visayas State College of Agriculture. The latter's program is supported by funds mainly from the International Development Research Centre (IDRC) of Canada. The impact is viewed not only with the improved varieties released but also, importantly, with the training of technical staff and strengthened networking with the national cooperative testing (NCT) stations, the established testing protocol, and with other state colleges and universities. The principal output of organized government efforts has been the development and release of improved varieties: seven from the Visayas State College of Agriculture, three from the IPB and two screened/released by the BPI from AVRDC lines.

Despite some weaknesses, the government seed program as a whole served as a catalyst to various crop improvement activities. Currently, private sector participation is existent only with the cash crops. The system with sweet potato and other root crops is basically government-heavy. Meanwhile, various informal farmer seed schemes prevail. At least 90% of the sweet potato growers have managed for years with this local system.

The seed production-distribution schemes tried in sweet potato have been well-intentioned but lack a systematic farmer-sensitive scheme and monitoring mechanism for farmer/user feedback or evaluation. Also, the top-bottom approach followed in HYV technology generation, failing to consider farming circumstances and nature of market, imposed a heavy toll - the non-adoption of the first VSP's. Reversals in approach had to face the stigma of the variety-market mismatch. The diversity of agroecological zones where sweet potato adapts and the peculiarities of user-based farmers' choices gave rise to several established good-performing local cultivars specific to an area. Partly, this has made the acceptance of the new high-yielding varieties relatively difficult. Only the improved varieties which approximated the "good-eating quality" criteria have been adopted by farmers. Sweet potato processing technologies which specifies certain physico-chemical characteristics offer market opportunities to the HYV's. Production and distribution of an improved variety, then, is a critical concern.

New challenges face the national program in developing an innovative simple, pragmatic and user-friendly seed production-distribution scheme.

INTRODUCTION

Of all the root crops, sweet potato has received the most research attention in the Philippines, even before the official creation of a national root crops center in 1976. Not only is it the most ubiquitous and easy-to-manage crop in various cropping systems, it has played an important role in subsistence as well as commercial farming. Scientists and researchers have pointed out its potential as a diverse source of human food, feed and other industrial uses. Thus, for more than a decade (1977-1988), sweet potato research at the Philippine Root Crop Research and Training Center (PRCRTC), the nationally-mandated center for root crops based at the Visayas State College of Agriculture (ViSCA), shared at least 85% of the overall root crops budget; 96% of this in varietal improvement. It was only recently that this breeding research received 32%. But the rest of the 68% also involved mostly sweet potatoes in such disciplines as pest management, postharvest, engineering, socio-economics, information/communication and extension (Palomar, M. K., 1989).

The International Development Research Centre (IDRC) of Canada has provided the majority support to all these research activities, about 80-85% of the rootcrops budget from 1977 up to the present. Most of the support was for breeding, multiplying and distributing improved varieties of sweet potato. While breeding activities and varietal breakthroughs are documented, very little is known of the extent to which the planting material has been multiplied, distributed and has reached the farmer. Since most of these projects have emphasized as clientele the small-scale and resource-poor farmers, it is of considerable interest to the donor institution, other collaborators as well as to the research and implementing agencies themselves to assess the extent to which the planting materials have reached their targeted beneficiaries, and the mechanism by which this was achieved.

Importance of the Study

Seed or planting material is an important input in the agricultural productive process whether of commercial value as in the case of traded seeds (i.e. cereals, other high valued crops) or home-grown or "asked" seeds (i.e. friends, neighbours). For years, the staple and major cash crops have received most research attention. As a natural consequence, official seed programs also concentrate at first on major and staple crops; less prominent crops like root crops have had extremely little research done. In the Philippines there is virtually no such study done on root crops other than that done on the white potato.

There is need to initiate seed production mechanisms study for sweet potato because availability of planting materials is an expressed constraint by farmers and the drive to expand the market via processing needs a systematic propagation-production link to sustain supply. The importance of a viable seed production program cannot be over emphasized if efforts to improve crop productivity are to be successful. Whether such goes through a developed official system or a simple, pragmatically designed, decentralized village-oriented seed system depends on specific conditions and criteria purposely considered. The system developed has to be anchored on an adequate understanding of existing systems practiced by farmers since they have long been in the art and, in their sense, "science" of seed propagation to sustain life for generations.

Objectives

The objectives of this study revolve around the generation and dissemination of information which will lead to the strengthening of systems to produce and distribute improved planting materials of sweet potato.

Specifically, the study aims:

1. To estimate the amount of improved planting material distributed, the number of growers using it and the areas planted; and

2. To describe and evaluate the alternative seed production and distribution mechanisms which have been tried.

This study also hopes to provide the national seed body useful information in working out a viable seed program for sweet potato in the country.

Methodology

The study adopts a combination of methods starting with a review of related studies and reports of verification trials and testing within the formal system.

A series of informal interviews of farmers during trips to sweet potato areas (i.e. Benguet, Northern Mindanao, Tarlac, Leyte) provided rich source of information. Key informants such as the sweet potato breeder research assistants involved in the trials and some members of the root crop technical working group were interviewed especially with the technicalities of varietal improvement and the testing schemes for (national) recommendation of a variety. Related studies and secondary data were quite helpful.

Formal surveys were undertaken in the provinces of Leyte and Samar (Eastern Visayas Region), Agusan (Eastern Mindanao), Catanduanes and Albay (Bicol Region). A total of 185 farmers were formally surveyed. The areas covered for both formal and informal surveys represent different cultivation systems and agroecological zones. The following presents the classification of areas:

1. Leyte (Jaro, Dulag, Alang-alang) - commercial lowland rainfed

2. Leyte (others: Baybay, Silago, Maasin, etc.) - semi-commercial/subsistence hilly/slopes

3. Catanduanes and Albay in Bicol Region - commercial/semi-commercial lowlands and undulating slopes/marginal
4. Samar (Pinabacdao, Sta. Rita and Calbiga) - semi-commercial/subsistence uplands/marginal

5. Agusan del Sur (Afga/Sibagat) - commercial undulating slopes

6. Benguet highlands - semi-commercial/subsistence (1500 masl)

These areas are presented in the map (Figure 1).

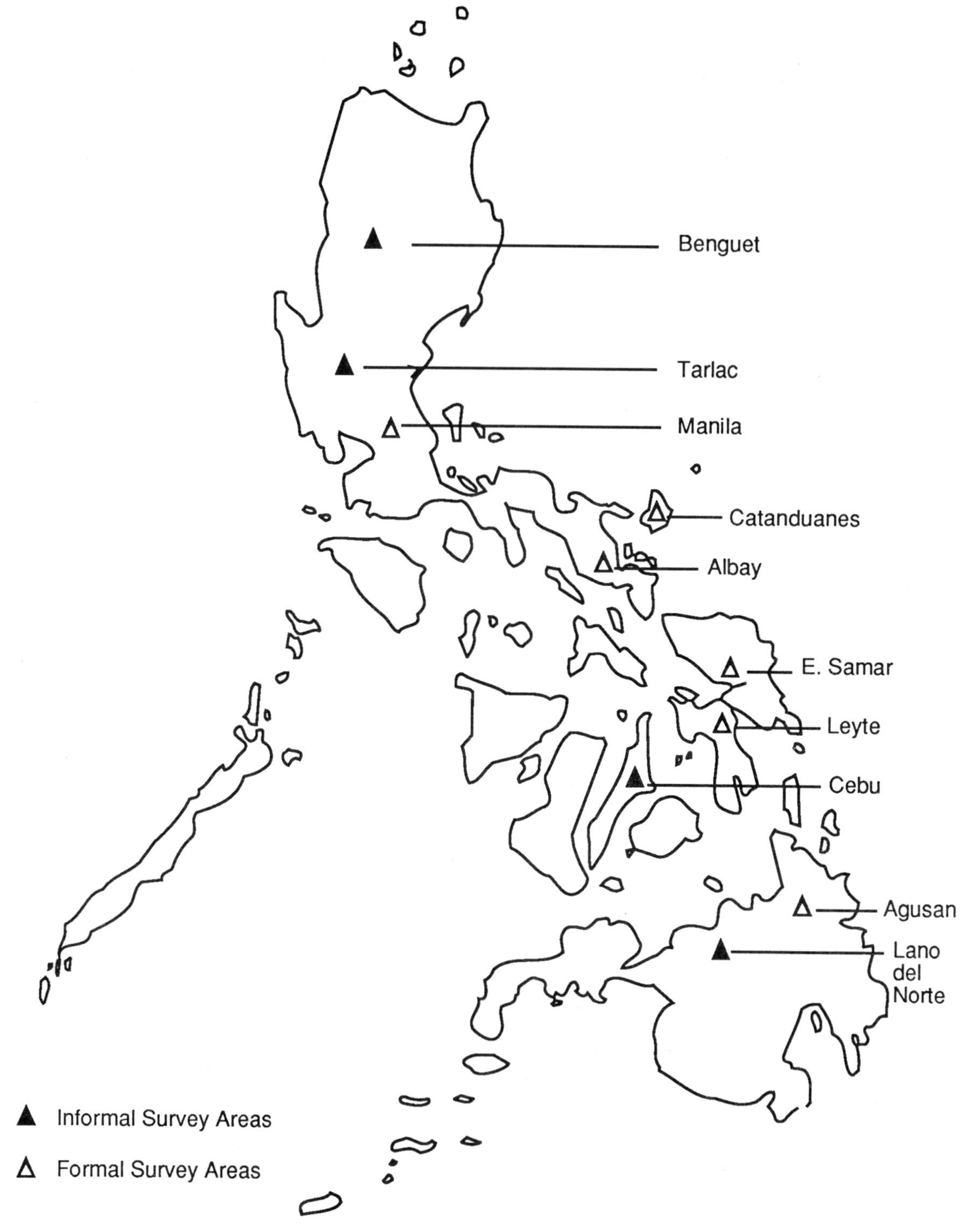

Figure 1 - Map of Survey Areas

Questionnaires were mailed to cooperating stations to determine the extent and nature of work on the HYV's. The total seed distribution was also estimated from the records of PRCRTC and the Department of Plant Breeding and Agricultural Botany at ViSCA.

BREEDING AND VARIETY SELECTION FOR SEED PRODUCTION

Background and Rationale

Sweet potato is an important part of various cropping systems in the country: in the cereals-dominated lowland, or in the mixed systems in the uplands and highlands, performing functions of sustenance and cash source. With the advent of simple technologies, scientists and researchers have pinned hopes on the potential of sweet potato as a nutritious processed food, basic feed ingredient and an industrial earner.

Also, sweet potato has been seen as a means of uplifting the lives of resource-poor farmers. But for it to do so needs a rationalized program since the fresh sweet potato market is very limited. The first two sweet potato regions, i.e., Eastern Visayas and Bicol have the highest incidence of poverty (50-60%) in the country and are typhoon paths. Clearly, sweet potato plays an important role as a cash supplement and famine-saving crop. About ninety-five percent of production is used as human food, mostly by boiling or the simple traditional processing (Figure 2). Per capita consumption, however, is low at 4-9 kg per year (FNRI, 1984). With a rice-based diet, the Filipino gets only about 5% of total starch intake from sweet potato (81% from rice, 9% from corn).

Figure 2. Sweet Potato Utilization and Starch Source in Diet

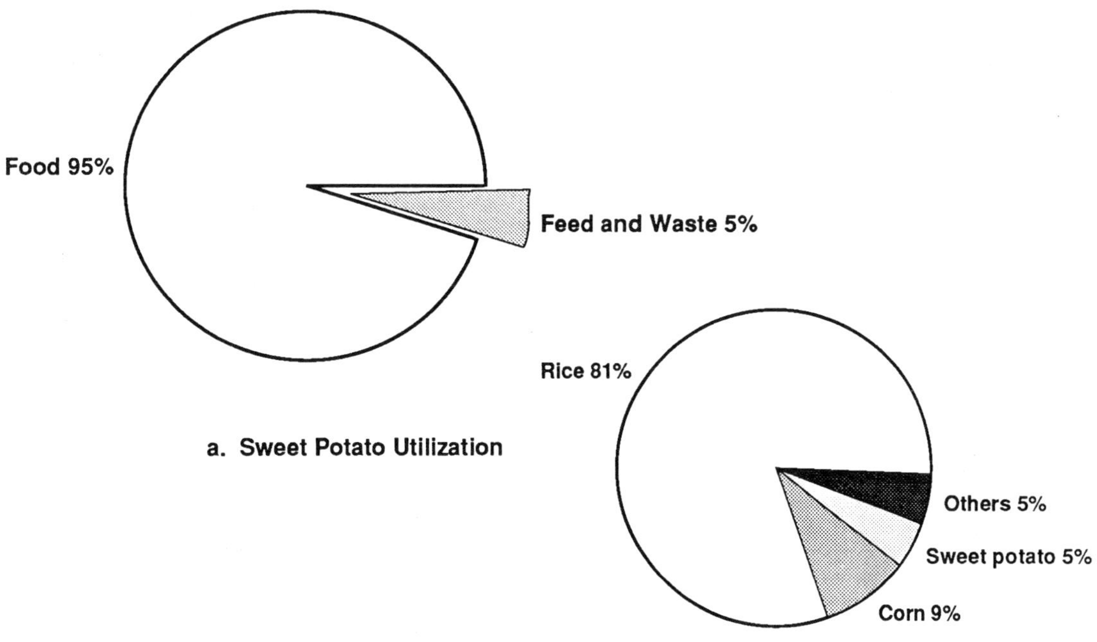

a. Sweet Potato Utilization

b. Starch Source in Diet

Partly due to this consumption pattern and partly due to national efforts to expand the grains program (especially upland rice and corn) and diversify agricultural production, the volume and area of sweet potato production showed a declining trend in the early eighties: from 235.8 thousand hectares in 1980 to 164.3 thousand hectares in 1987 and a production volume of 1.05 million metric tons to 0.84 million metric tons, respectively (Table 1). With the average yield relatively constant (i.e. about 4.7 tons/hectare) for the period, reduction in area due to substitution or increased cropping intensity of another crop (e.g. rainfed rice, corn) offers a plausible reason.

Table 1. Area Planted and Volume of Production for Sweet Potato

	Area Planted (hectares)	Volume of Production (000's MT)
1978	227.6	1,037.0
1979	238.0	1,122.9
1980	235.8	1,047.8
1981	220.9	1,010.3
1982	209.3	1,037.6
1983	174.7	801.5
1984	170.1	820.3
1985	164.3	777.1
1986	164.8	800.6
1987	164.3	843.7

Source: Bureau of Agricultural Statistics, Department of Agriculture, Philippines.

With the existing constraints of a limited fresh roots market, development research on sweet potato follows a market expansion-diversification scheme to stimulate farmers to produce. Processing technologies have been developed including various uses of sweet potato flour, beverage, catsup, delicious SP, jam, fruit-like products, naturally fermented soy sauce, etc. Some are already in their pilot stages. The viability of processing technology, however, is premised on reasonable costs of inputs (via high yielding varieties) but that which gives the farmer sufficient returns, in addition to improving varieties for table use. Such becomes the rationale for varietal improvement. And, concomitantly, a system of propagating and distributing planting materials for sustainability of the whole process.

BREEDING AND VARIETY SELECTION

Historical Sketch: Overall Sweet Potato Breeding Work

Before the 70's, sweet potato breeding work was rather rare and limited. The recorded pioneering breeding program of Mendiola (1921) which produced fancy strains of sweet potato was short-lived. Other sweet potato research work (i.e. occasional studies in varietal evaluation) of some agricultural colleges were also limited. This was highlighted by the release of a new sweet potato variety, BNAS 51, which became a standard check in later sweet potato experiments.

A financial research grant from the National Science Development Board in root crop research spurred attempts at sweet potato improvement in the University of the Philippines at Los Banos (UPLB) in the early 70's (Carpena, 1975). Since then, the interest on root crop research began to build-up with the creation of the National Root Crop Research Center in 1976 (now PRCRTC). PCARRD (Philippine Council for Agriculture and Resources Research Development) granted the Visayas State College of Agriculture research funds for the collection and evaluation of the local and introduced root crop varieties. In the same year, the International Development Research Centre (IDRC) of Ottawa, Canada gave PRCRTC a grant for multidisciplinary research on root crops.

From then on, three sources of new sweet potato entries have been identified: from the (1) Philippine Root Crop Research and Training Center, ViSCA; (2) The Institute of Plant Breeding (IPB), UP Los Banos research group; and (3) the Bureau of Plant Industry (BPI) where selections from breeding lines of the Asian Vegetable Research and Development Center (AVRDC) were entered.

PRCRTC-ViSCA Breeding

The 1975 PCARRD grant (Project #259, Saladaga, 1976) enabled the collection (for a germplasm pool) and evaluation of introduced and local cultivars to identify those with promising traits for release as varieties or for use as parents in hybridization. In 1979, the project identified two varieties for mass production and distribution to farmers: BNAS 51 and San Isidro.

The sweet potato varietal improvement (first phase) was anchored on the objective of producing a variety with most, if not all, the traits desired by farmers and consumers. These traits include high root yield, high dry matter content, early maturing, resistance to weevil and other pests and diseases (e.g. scab, leaf spot), high protein content, acceptable weight loss in storage and, in general, good eating quality. The critical variables which have been considered for national recommendation of a variety by the root crop technical working group of the Philippine Seed Board are root yield, resistance to scab and physico-chemical properties as dry matter, starch, sugar and protein content.

To date seven new varieties were released for national recommendation from the breeding lines developed at PRCRTC, ViSCA (i.e. VSP 1 - VSP 7). These were based on two-season (wet and dry) results of five regional trials in different cooperating stations in the country.

In 1989, IDRC approved an integrated root crop development program where sweet potato varietal improvement is an important part. Unlike the breeding program in the past, the method currently followed reflects the bottom-up approach, integrating user-orientation in the process of generating and evaluating technologies (i.e. HYV, practices, etc.). Another bent is the giving of priority to small, subsistence and semi-commercial farmers where the agroecological zones are not the first class lowland relatively fertile zones characteristic of commercial sweet potato farmers in the country. The latter constitute only about 5-8% of total area devoted to sweet potato where they are grown as cash crops with net value added even better than either rice or corn.

Genetic Basis for Improvement

The polycross breeding technique was applied and modified to suit the needs of sweet potato for increasing variabilities. This technique allowed the production of numerous recombinant genotypes over a relatively shorter period, thus overcoming the problem of low seed set common with controlled biparental crosses. Rapid evaluation and screening procedures had to accompany this technique for efficient results.

The first three newly released sweet potato varieties, VSP-1, VSP-2, VSP-3 and succeeding releases had been genetically improved through the polycross technique. Among the parents in the polycrosses were native cultivars adapted to Philippine conditions. Recurrent selection for these traits in subsequent progenies had increased the frequency of genes for adaptability to Philippine conditions and resistance to disease (i.e. sweet potato scab). Meanwhile, the polycross technique ensured the maintenance of the highly heterozygous genetic nature of these progenies among which were selected breeding lines later renamed VSP-1, VSP-2 and VSP-3. Later releases had incorporated genes controlling traits desired by subsistence farmers and consumers, i.e. high dry matter content and long vines that produce roots along the nodes of the crawling vines for the staggered harvesting practices (Saladaga, FA., personal communication).

The nationally recommended sweet potato varieties developed from the varietal improvement program of ViSCA are presented in table 2.

Table 2. Matrix of Recommended Varieties and Characteristics of the VSP Varieties

Source: Villamayor, Federico G., PRCRTC, ViSCA

Characteristics	VSP1	VSP2	VSP3	VSP4	VSP5	VSP6
Morphological						
Root skin color	red	orange w/purple spots	red	white	red	red
Root flesh color	orange	orange, purple spotting	yellow	yellow with orange spots	purple	light yellow
Plant type	spreading	spreading	spreading	spreading	spreading	spreading
Mature leaf color	green	green	green	green	green	green
Petiole pigmentation	moderately purple	purple	green with purple tip	moderately purple	green with purple tip	green with purple tip
Agronomic						
Yield potential t/ha)	21	19	17	16	17	19
Harvest age (days)	90-100	90-110	100-120	90-100	90-100	100-120
Susceptibility to:						
Weevi	moderate	high	moderate	high	moderate	moderate
Scab	moderate	moderate	high	moderate	moderate	moderate
Tolerance to:						
Poor soils	not	not	moderate	not	not	n.d
Shade	not	moderate	high	moderate	moderate	moderate
Drought	not	not	high	high	high	high
Storage						
Weight Loss	high	high	moderate	moderate	high	n.d.
Sprouting	very low	low	very high	good?	low	n.d.
Rotting	low	very high	very low	low	high	
Physico-chemical						
DM content	26.5	33.5	34.1	34	32	37
Startch	56-73	59-75	65-83	65-82	62-82	60-80
Sugar	13.6	13.4	7.1	6.6	6.4	9.4
Protein	2.6	2.7	1.9	2.1	1.4	1.7
Recommended use:	Fo/Fe	Fo/Fe	Fo/Fe	Fo/Fe	Fo/Fe	Fo/Fe

Fo - food Fe - Feed N.D. - no data

Source: Villamayor, F.G., PRCRTC, VISCA

High yield and early maturity are the main criteria considered for national recommendation. VSP1 has the added advantage of having a high beta-carotene content and is recommended for nutrition-rich food product. VSP3, 4 and 6 are the varieties which closely fit the type for table use. VSP5 could be a cheap substitute of yam in food processing.

The profitability analyses done was mostly a comparison of the VSP's with yield as the variable factor based on experimental conditions. This was rather a limited approach since demand is a critical factor which was rather difficult to show with the first VSP's.

Two-season (wet and dry) two to three year trials in the different collaborating station are undertaken on the average after a variety has been identified in the originating institution.

III. MECHANISMS OF SEED PRODUCTION AND DISSEMINATION

The Formal Seed System: Brief Historical Background with Emphasis on Sweet Potato

May, 1954 marked the first annual meeting of the cooperative seed improvement group initiated by the Bureau of Plant Industry, the University of the Philippines College of Agriculture and Department of Agriculture to form a body to pass on or approve varieties before any crop variety will be increased. The result was the creation of the Philippine Seed Board created by Special Order No. 590 Series 1955.

The primary concern for food security figured not the immediate agenda to focus on the two staple crops, rice and corn. The varietal improvement criteria were high yield, good eating quality and desirable agronomic characteristics.

Beans and vegetables were included in 1956. In 1969, specific Working Groups were designated to work on different aspects of seed improvement such as cultural practices, fertilization and screening tests for disease resistance. 1969 also marked the inclusion of root crops especially sweet potato in the Seed Board's Improvement program.

Various revisions were made to accommodate new crops and add on critical characteristics for varietal improvement. These lead to the formation of several technical committees in 1982: varietal improvement, seed production, seed distribution and seed storage, seed certification, seed standardization, promotion and extension.

In 1982-83, the root crop technical working group requested to establish specific guidelines to consider in the conduct of various tests needed in varietal evaluation and to include cultivars grown by farmers. The relatively low adoption rate of recommended varieties despite high yields prompted the suggestion to include eating quality and physicochemical analysis of the selection for release as well as improved seed production and distribution for crops other than rice and corn. The membership of the root crops technical working group expanded to include more state colleges and universities and BPI stations for testing and evaluation. Regional recommendation of varieties was brought out of diverse agroecological conditions and thus, varietal adaptability. Part of the discussion was the suggestion and that data on farmers field trials, not only those of experiment stations, be part of evaluation of the varieties.

Later, sweet potato varieties recommended were more attuned to the preferences of consumers. The 1988 meeting called for the review of performance of recommended varieties and suggested that a small committee from the root crops technical working group monitor and evaluate released varieties and gather information on utilization and acceptability of the varieties in coordination with the Department of Agriculture field offices. No report has been submitted yet.

In general, the varietal evaluation scheme for crops is presented below. The complete flow including certification is present only for the major grains (i.e. rice and corn) and some cash crops (i.e. some vegetables and exportable fruits). The participation of the private sector in the multiplication of certified seeds can also be seen with these crops.

Figure 2. General Crop Varietal Evaluation Flow Scheme

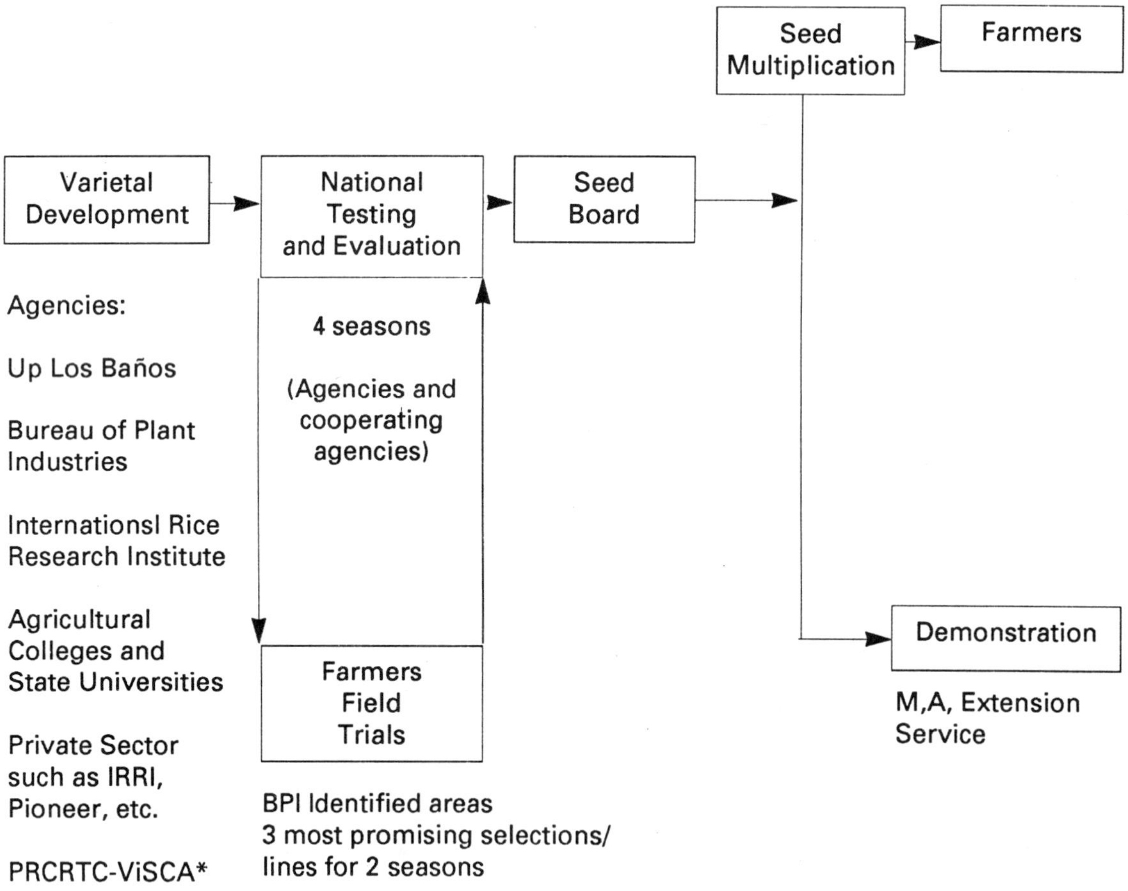

1/ Adopted from the Phillipine Seed Board files.

* national center for sweet potato/rootcrops

Figure 3. Testing Protocol

A. Rice, Corn, Some Vegetables and Legumes

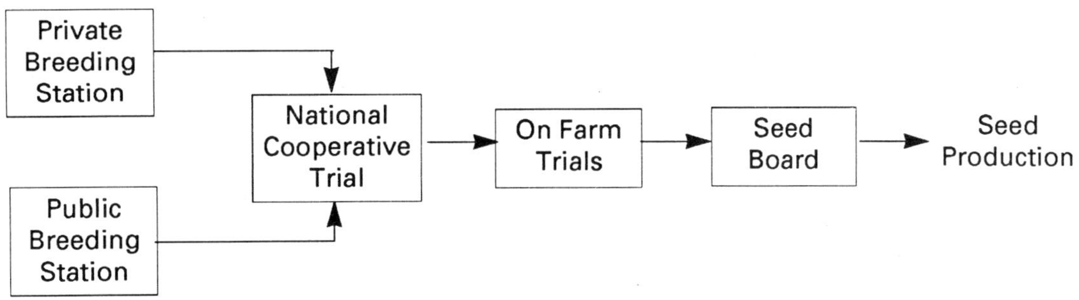

Alternative:

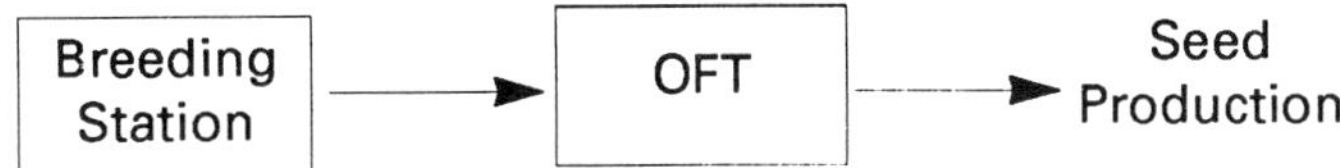

B. Fruits, Plantation Crops

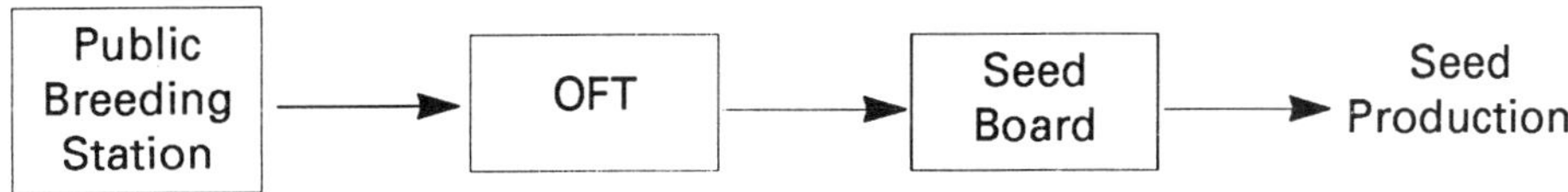

C. Sweet Potato

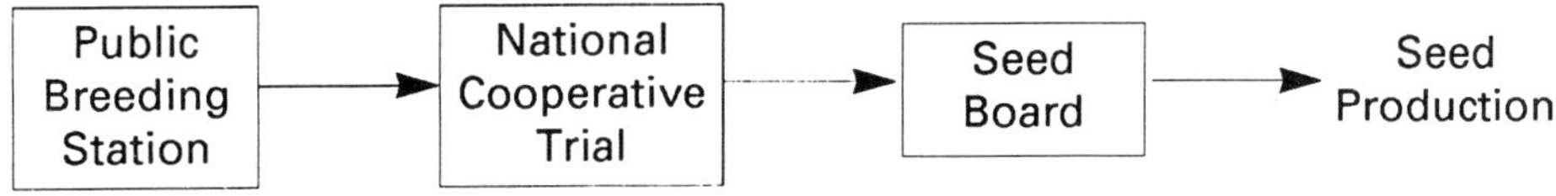

Note: A and B adopted from E.T. Rasco, Jr.

Figure 4. Seed Production and Distribution

A. Rice, Corn, Some Vegetables and Legumes

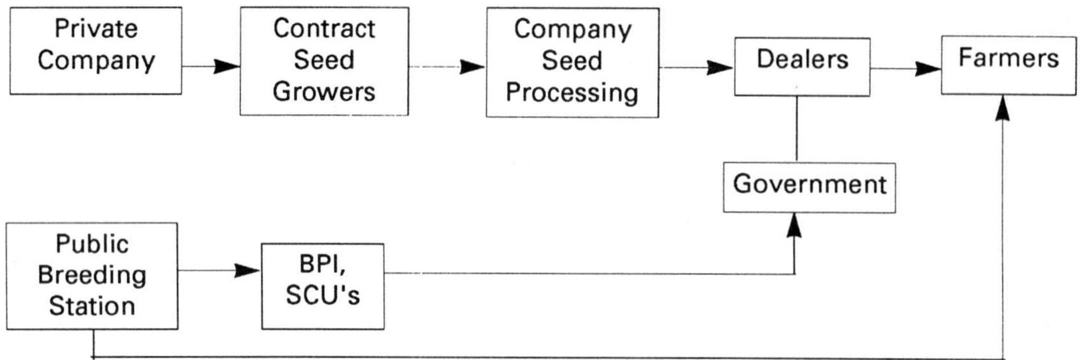

B. Fruits, Plantation Crops

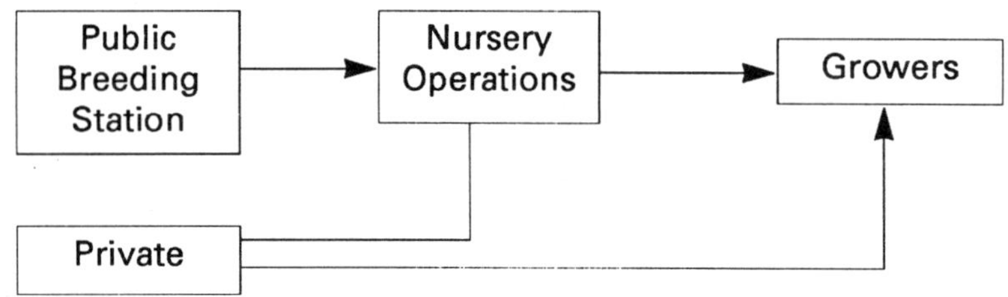

C. Sweet Potato/Root Crops

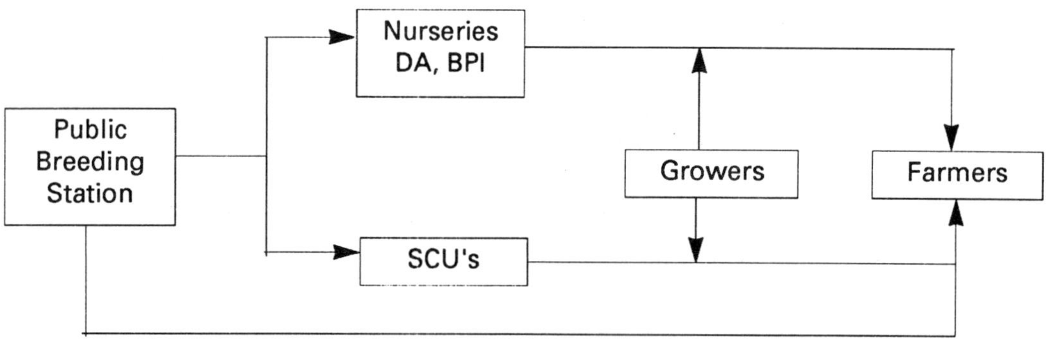

It is only in the grains, vegetables, some legumes, fruits and plantation crops where participation of the private sector is evident for seed production and marketing. The sweet potato/root crops system is government-heavy for the whole breeding-evaluation-distribution flow. Recent developments call for a greater farmer and private sector participation but this is still in the developmental phase.

Distribution through the National Cooperative Testing Network

This is a network of thirteen experiment stations of the different regions throughout the country where crops, including root crops, are tested and evaluated for national or regional recommendation. These consist mostly of the Bureau of Plant Industry of the Department of Agriculture stations and some state colleges or research centers.

The NCT stations also serve as propagation centers where nearby farmers or agencies can request. The estimate of planting material distributed is very conservative since most stations could not respond due to lack of monitoring system.

The PRCRTC-ViSCA Production-Dissemination Program

The components include the breeder's trials, training, extension projects and varietal testing outside the National Cooperative Testing (NCT) Network.

Breeder's trials. On-farm trials to verify experiment station results and for propagation have since the start been an informal means of disseminating the VSP's. These were mostly researcher-managed and funded by government or an external donor (to a large extent) through Center projects. During the first phase of the sweet potato varietal improvement program which produced the first two VSP's (i.e. VSP 1-2), trials were done in the sweet potato growing areas in Leyte which were mostly lowland farms. One or two cooperators (i.e. either a farmer leader or a MAO technician) were selected for each site on the basis of contact and initial interest. The results showed the high-yielding and early maturing characteristics of the VSP's which triggered other farmers to get cuttings from the cooperators or bought these at P0.03-0.05 per cutting from ViSCA or through the DA technicians. At least 300 farmers in these areas were served in the first phase. However, the market became an acid test: the orange, moist-fleshed sweet potato did not fit the fresh market demand. The ViSCA-developed delicious SP processing in a nearby city (i.e. Tacloban) saved the fate of the 1st VSP's - the daily need was at least 500 kg of VSP1. The untimely demise of the operations (i.e. non-recovery of the factory after fire) greatly lessened the stimuli to grow the VSP's. Farmers simply maintained a small portion for home use or just as not to lose the variety. Scab disease in 1987 caused a total eradication of the varieties.

Currently, the breeder's trials (i.e. researcher-supervised/farmer-managed) are intensified in Matalom, Leyte - a location which approximates the agroecological zones of subsistence and semi-commercial root crop farming (i.e. uplands, marginal lands). Pinabacdao, Samar, the first pilot site of the new IDRC program in the Philippines also carry on these trials (i.e. on-site station and farmers' fields).

Training: Training served as the main vehicle in the dissemination of the VSP's. Participant/recipients were mostly farmers, technicians and some private entrepreneurs and non-government organizations (NGO's). The production training which started in 1985 consisted mainly of introducing the varieties and the management practices recommended. Usually, each participant is given free 20-50 cuttings for each variety. There was a high mortality rate of the cuttings when training was Center-based due to distance. Most on-site training did have better results.

Extension projects: The transfer of sweet potato technologies such as storage and food processing usually integrates seed distribution. This was done by the project/study leaders themselves or in collaboration with local agriculture technician. These include several towns in Leyte and Southern Leyte (e.g. Silago, Maasin, Bontoc, Malitbog, Padre Burgos, Capoocan, Baybay and Dulag).

Outside NCT Trials. PRCRTC runs a project of testing the adaptability of recommended varieties and promising accessions in other locations in collaboration with state colleges and universities not covered by the National Cooperative Testing (NCT) Network. These SCU's include the Catanduanes State College (CSC), Tarlac College of Agriculture (TCA), Pangasinan State University (PSU), University of Southern Mindanao (USM), Central Mindanao University (CMU), Isabela State University (ISU), Don Mariano Marcos State University (DMMSU), Northern Mindanao Institute of Science and Technology (NORMISSIST) and Misamis Oriental State College of Agriculture and Technology (MOSCAT). The trials (researcher-managed) consisted of single rows to replicated trials at their experiment stations or farmers' fields which also served as propagation areas and distribution points.

Farmers' Field Day/Exhibits. The farmers' field day is one of the highlights of the College anniversary where the root crops center traditionally takes this as an opportunity to disseminate information on technologies including the distribution of planting materials.

National (e.g. Philtrade fair) and local (e.g. town fiesta, anniversaries) exhibits provide yet other venues for distribution, usually free.

The Department of Agriculture Extension Activities

The role of the DA in planting material distribution works in any or a combination of these ways: (1) as an outgrowth of the experiment station trials in cases where DA-BPI is an NCT member, (2) as part of an emergency response to flood or typhoons, or (3) as part of DA-led or linked special projects involving root crops.

The first one is discussed in the NCT Network. Sweet potatoes as stop-gap to famine or lack of food is a familiar measure in areas seriously hit by typhoons or floods like Region V (Bicol) and Region VIII (Eastern Visayas). This was especially so in the mid-80's where the VSP's were first popularized from the recommendation of the Philippine Seed Board. Local DA's secured the improved varieties from ViSCA. The areas served were northern towns of Leyte, Eastern Samar, Catanduanes and Albay in Bicol.

No estimates could be reached and these were not recorded nor monitored. Also, the DA usually requested cuttings from ViSCA for flood or typhoon victim propagation and this must be recorded at the ViSCA level.

Private Sector Projects; Informal Exchange Between Farmers

Enterprising farmers who initially got their cuttings from ViSCA did propagate in their own farms and maintained those varieties mostly preferred. Some farmers in Cebu, Leyte, Cotabato, Pampanga, Tarlac, Albay and Catanduanes were traced. Not less than half a million cuttings were distributed in a smaller (a few hundreds per time) or bigger scale (by hectares per time). Noteworthy to mention are private sector propagation initiatives in Cebu, Pampanga - Tarlac and Cotabato - Davao.

The Informal System: Farmer Seed Systems

At least 90% of sweet potato growing thrives through the informal distribution system. Farmers, whether subsistence or commercial, are in their own right natural breeders and nursery operators. Their main considerations for selection are user's preferences and fit in their cropping systems. Yield, early maturity and resistance are important, though not necessarily the first priority. Their method of propagation and sources of planting material are dependent on their scale of operation, space availability and cropping pattern; the latter being a function both of the environment and the value and relative importance of the crop to the household demands for consumption and other socio-economic needs.

The various farmer seed systems identified by using both focused formal and informal surveys (in Luzon) are described below.

Farmer/barter exchange scheme. Farmers within or in neighbouring localities exchange planting materials of varieties known to perform better and introduce them into their own fields. Over the years, farmers have been undertaking a "breeding" scheme and select for the more stable and preferred cultivar. In the traditionally commercial sweet potato areas, the farmer-traders carry some planting materials to other places and exchange them with popular cultivars in other areas. Another point of exchange is when two areas have different growing seasons and both widely popularize a certain cultivar. When planting material propagated is not enough, farmers seek for the materials in another area about to harvest. And the latter's farmers do the same in the reverse of the season.

Plot propagation/transplant scheme. This is practised mostly by commercial farmers in cropping systems where sweet potato is rotated with rice (rainfed), corn, vegetables or legumes. A portion of the sweet potato field is left after harvest as source of planting materials for the next season. Since harvest is timed at the onset of the dry season, the transplanting (for a wider plot propagation estimated by the farmer as enough for his intended area of production) is done when the rain starts to come or when adequate moisture is expected. The area prepared is either the adjacent plot or another plot (usually conveniently nearby) and ranges in size from 10-200 m^2. In bigger production areas (e.g. Paniqui, Gerona, et.al. of Tarlac province) transplanting is usually twice or thrice (starting in May-July) and the plot is a clean rectangular etching in the middle or side of the rice field (rainfed) which transforms into a sweet potato field from November to April. In the sweet potato-corn-vegetables/legumes commercial cropping systems (e.g. Agusan, Leyte, Lanao del Norte, Batangas, etc.), the plots are relatively smaller than those in Tarlac ranging from 10-50 m^2 and usually are in the peripheries or nearby the cornfield.

Backyard/Peripheral Patches. In subsistence and semi-commercial sweet potato production where farm sizes are small (0.25 or less), the intensive use of land to provide for sustenance makes irrelevant to the farmer the provision for propagation plots. The best means of preserving planting materials for the next season is to let the sweet potato grow in the peripheries of the next crop or maintain them in backyard patches or home gardens. A conservative estimate of 80% of sweet potato-growing farmers apply this system.

Direct Planting Scheme. In commercial and semi-commercial sweet potato producing areas, where the distribution of rainfall is relatively uniform and there is no pronounced dry spell, sweet potato is directly planted to already prepared growing areas. A parcel can have two cropping intensities for SP. With 2-3 parcels per farmer, sweet potato is directly planted to another parcel when a farmer needs a corn or a legume for a third crop. The rotation is repeated in the other parcels. This system provides the farmer the produce he needs for home and market and preserves the fertility of the soil. In this system, sweet potato is grown all year round (e.g. Agusan and Lanao del Norte).

Cooler/Shady Location Propagation. In areas with pronounced dry spell or drought-prone areas, farmers preserve planting material by propagating in small patches (5-10 m^2) under the shade of trees (e.g. coconut, etc.) or in cooler upland areas.

Farming Circumstances

General Characteristics. The areas surveyed produce sweet potato in a semi-commercial to commercial scale with the Butuan-Sibagat producing area bend (Agusan) as mainly commercial. Sweet potato is planted to 25-30% of total farm area. The rest are shared either as intercrop or rotating crop with crops such as rice, corn, vegetables and with supplemental crops such as bananas and other root crops. Coconuts are important cash sources. In the Bicol region, sweet potato is mainly a monocrop. The cropping patterns adopted by farmers is a clear indication of the farmers' ability to cope with the given physical environment, farming household needs, resource constraints and market potentials. Sweet potato is largely an important supplemental cash crop in these areas.

Sources of Planting Materials. Most farmers depend on their own farms (70-92%) for the next season's requirement of planting materials. Neighbouring farmers are also important sources. A few farmers in Leyte and Samar (4-11%, respectively) bought planting materials from P0.07-0.15 per cutting. In Albay, the BIADS, a cooperative of farmers, became an important source of planting materials at P0.10 per cutting.

Other farming conditions are summarized in the following table 3.

Table 3. Summary of Background Characteristics in Survey Areas and of SP Planting Material Propagation.

	LEYTE n=24S	AMAR n=61	AGUSAN n=26	CATANDUANES n=6	ALBAY n=10
Major Crops:	rice, coconut, vegetables, corn	rice (upland) corn, vegetables	bananas, corn, rice, coconuts	rice, coconuts	rice, coconut, corn
Supplemental Crops:	bananas, root crops (sweet potato)	bananas, sweet potato	vegetables sweet potato pineapple	root crops (Mainly sweet potato	sweet potato
Root Crops:	sweet potato gabi cassava ubi	sweet potato cassava gabi ubi	sweet potato gabi cassava yautia ubi	sweet potato cassava	sweet potato cassava
Type of SP Market Orientation:	semi-commercial	semi-commercial/ subsistence	commercial	semi-comercial/ commercial	semi-commercial/ commercial
Ave. Farm Size:	1.0 - 1.51	0 - 1.51	0 - 1.51	0 - 1.51	.0 - 1.5
Ave. SP Farm Size:	0.25 - 0.50.	25 - 0.50.	25 - 0.50.	25 - 0.50.	25 - 0.5
Years in Farming:	21	23	20	27	21
Cropping System: (Main)	Intercropping 58%, Crop rotation 29	Intercropping (57) Crop rotation (30)	Crop rotation (27) Intercropping (50) Monocropping (19)	Monocropping (mainly)	Monocropping (mainly)
Others:	Monocropping 8 Mixed 4	Mixed (7) Monocropping (5)	Mixed (4)	Intercropping, crop rotation	Intercropping
Characteristics of SP Farms:					
- Soil type	clay loam (100)	Clay loam (74) Sandy loam (26)	Clay loam (73) Sandy (23) Silt loam (4)	Clay loam (83) Sandy (17)	Sandy loam (50) Clay loam (50)
- Topograph	flat (100) hilly/rolling)	flat (26 hilly/rolling (74)	flat to rolling (100)	flat (17) rolling/hilly (83)	flat to rolling (100)
- Climate	distinct dry and wet	distinct dry and wet	uniform rainfall	distinct dry and wet season	distinct dry and wet season
Sources of Planting Materials:					
- own farm	92	79	73	70	76
- neighbours	8	21	15	45	57
- others	-	-	11	17 (landlord)-	35 (BIADS)
Bought Planting Materials:					
- Yes	4	11	-	-	50 fr.(BIADS)
- No	96	89	100	100	-
Price, if buy:	P0.10 per cutting	P0.07-0.15 per cutting	-	-	P0.10 per cutting
System of Planting Material Propagation:	* Cuttings are let to stay for about 3 days in a cool, shady area before planting (71%)	* Direct planting after harvest (59%) * Plant cuttings on prepared plot before transplanting to a bigger area (41%)	* Direct planting (100%)	* Direct planting * Certain portion of \field is left unharvested as source of planting materials * Plant in peripheries of corn fields backyard plots of $10m^2$ after last harvest (common practice)	* Not harvest a portion of field - let grow for propagation * Plant in peripheries * Propagate in plots under coconut ($10x10\ m^2$ or $5x5\ m^2$)
Plot sizes, if practiced (most common)	-	$2 \times 4\ m^2$ $2 \times 2\ m^2$ usually backyard plots	-	$10\ m^2$	$5 \times 5\ m^2$ $10 \times 10\ m^2$

Tried Improvements to Limiting Factors The earlier informal and formal feedback on the non-acceptance of the first moist varieties spurred attempts to find uses and capitalize on the positive aspects of the new varieties and, in general, to work approaches at improving development, production and dissemination of new varieties.

Sweet Potato Processing. Food technologists at ViSCA worked around the relatively high beta-carotene content and attractive color of VSP1. The focus of developing nutritious products from sweet potato became the fit to this highest-yielding variety among the VSP's. Efforts resulted in the making of the sweet potato beverage (currently adopted by a big corporation), delicious SP (similar to dried mango), catsup, jam and SP chips and sticks (snack products). The demand for VSP1 has not been served and definitely, a seed production-distribution scheme should be in order. The creation of a demand for this earlier rejected variety via processing has niched out the critical concern on market.

The Advent of Social Science. The recognition of the importance of social science in the whole chain of technology generation to transfer led to the creation of the socioeconomic section in the last quarter of 1987 at the root crops center. Being ultimately skeletal, it works with affiliates from other social science departments at ViSCA. The minimal advantage is that a core staff works fully on root crops. With this trend, social science approaches has been heightened and integrated in various projects.

Seed Production and Distribution Integrated in Pilot Projects

The use-specificity of the new varieties demands the inclusion of production and distribution of planting materials in processing pilot projects. A case in point is the try-out scheme followed in the naturally fermented sweet potato soy sauce project. Utilizing the relative strength and reasonably good linkage between the local agriculture extension agent and the beneficiary farmer group, a few hundreds of planting material and a simple use-awareness of the varieties through various consultation meetings were all it took for the project team on this component of piloting. All the work was done basically by the farmer group with the assistance of the local technician assigned - from the selection of propagation areas and demo plots, propagation and linkage with other farmers who needed the materials. Farmers, too, set their own terms. This important component of the project was largely eased out of the team who had more pressing technical matters to face. Highlighted in this particular case is the importance of the capability of extension agents and farmer groups in working out a village-based scheme.

Rationalization of Training and Extension Program

The need to be relevant and sensitive to users' circumstances has led to the strengthening of the rootcrops training and extension program. Salient features include carefully designed on-farm trials, simplified training for farmers and would-be processors, use of appropriate dissemination media, strengthened linkages with local groups and a built-in projects' monitoring scheme. These are currently seriously considered for adequate and continuous Center support, not co-terminus with projects as in most cases in the past. A critical concern is the provision for support (defined as to nature of project) during the transition phase of period-completed pilot cases. The sustainability question could hinge on other support services identified in the process. Referrals and further strengthening of linkages may be necessary.

IV. ASSESSMENT OF DIFFUSION AND ADOPTION

The attempt to evaluate how fast and to what extent the varieties have been adopted is relatively difficult particularly in a system where monitoring is not built-in. Estimates of areas grown and the quantity of planting materials distributed are largely drawn from records and surveys, both informal and formal (n=58). Survey results are summarized in the table at end of section.

The Diffusion Process

To the cooperators, the ViSCA project/study leaders in their trials or extension projects were the main source of information (60%) as well as planting materials (60%). The agriculture technicians are also important vehicles of information relay in all cases. The informal exchange among farmers proved effective. Fifty-eight percent of the non-cooperator/adopters got the technology through this process.

The cuttings were mostly given free and in some cases at a cost of P0.03-0.10 per cutting.

Extent of Adoption of VSP's

Respondent cooperators grow on the average 1.4 hectares of sweet potato, about 42% of total farm size cultivated. Fifty percent of the sweet potato area is planted with VSP's and the rest on the popular local cultivar. For non-cooperator/adopters, 0.7 hectare or 21% of total farm size is grown with sweet potato. About 43% of the sweet potato area is devoted to VSP's. A conservative nation-wide estimate of area seeded to the VSP's is a little less than 1% of total sweet potato area, most are commercial or semi-commercial.

Most VSP's tried were the first four released varieties, VSP 1-4 with VSP 1 and 2 the least popular in Leyte area where commercial growing serves mostly the fresh food-traditional use food market. The opposite is true in Bicol where sweet potato catsup processing is tried out; VSP 1 and 2 stood out. VSP 4 is already gaining wider market acceptance, and thus, farmer adoption is most areas where introduced. Dulag, the only town in Leyte known for commercial sweet potato production, has widely accepted VSP 4 with traders gaining repeat orders from city buyers. In other Leyte towns, Pangasinan, Pampanga, Bulacan and Bicol, VSP 3, 4 and 6 are also gaining acceptability. These varieties approximate the eating qualities desired by consumers. VSP 3 has been especially recommended for sweet potato-cassava-feedmill project (Pangasinan). Some farmers prefer VSP 1 as feed to pigs because of their relatively high vitamin A content. In the southern towns of Leyte VSP 3 and 4 are also accepted together with the earlier ViSCA varieties introduced, the BNAS-51 and V2-42.

The collaborating stations in Mindanao have been propagating and distributing cuttings in the provinces of Bukidnon, Misamis Oriental, Agusan, Cotabato and Zamboanga. No adequate farmer feedback was given but preliminary reports suggest positive gains for VSP 3 and 4. The same is true for Cebu in Central Visayas where an enterprising farmer has kept sufficient supply of the VSP's. On the whole, VSP 3 and 4 fared well in areas where they had been tried.

The current increasing demand for VSP 1 is due to the transfer of processing technologies such as the sweet potato beverage in Pampanga (which links with the commercial sweet potato farmers in Paniqui, Tarlac - Central Luzon region) and the sweet potato catsup processing in Catanduanes and Albay (Bicol Region).

Others have learned to make catsup with other cultivars by simply playing with the formulation and color.

The VSP's in general did not perform well in the highlands of Benguet where high-yielding local cultivars dominate. The VSP's were products of lowland breeding. Most sweet potato farms are typically upland. This lack of agroecological fit was a source of risk and frustration.

In general, the acceptability of the VSP's stood in direct correlation to use and market acceptability. While farmers were impressed by their yields and early maturity, the new varieties have to stand the test of the market and users. The negative image of the VSP's which resulted with the first introduced moist types is now gradually phased out as awareness of different varietal types, their respective characteristics and uses are being emphasized in extension and distribution activities.

Also, the existence of good performing local cultivars in the different sweet potato areas made the inroads to acceptance of the VSP's relatively difficult. Examples are shown below:

Area	Cultivar	Maturity	Yield (t/ha)	Eating Quality
Leyte	Miracle	3-4	12-14	Good
	Karingkit	5-7	6-8	V. good
	Siete Flores	5-7	6-8	V. good
Benguet	Kalbooy	5-7	20-25	V. good
Tarlac	Bureau	2 1/2-3	16-18	V. good
	Taiwan	3-4	16-20	V. good
Agusan del Sur	Katimpa	4-5	8-12	V. good
	Kasima	5-6	8-12	V. good
	Senorita	4-5	8-12	Good (Chinese)
Iligan	Makapuling	5-6	14-18	V. good
	Imelda	5-6	14-18	V. good
	Chinese	5-6	14-16	Good (Chinese)
Batangas	Sinuksok	4-5	10-12	Good
Quezon	Taiwan	3-4	14-16	V. good
	Miracle	3-4	10-14	Good
Bicol	Tres Colores	5-6	10-12	V. good
	Sinimet	5-6	10-12	V. good

Sixty percent of the cooperators are still growing the VSP's compared to thirty-eight percent of the non-cooperators/adopters. The main reason for not growing is the dearth of planting materials due to extreme weather conditions (drought or severe rain). Farmers complain that the VSP's are sensitive and cannot withstand extreme conditions. They report, too, that continuous cropping of the VSP's without substantial inputs yield lesser output. The creeping local cultivars can be conveniently grown for longer period and, thus, less work-intensive.

Other constraints in the adoption process include lack of resources to purchase required inputs (i.e. fertilizer), lack of adequate information of the HYV technology and the unavailability of technical assistance when needed.

Estimates of Material Distributed

Based on records at ViSCA and responses from the stations and collaborating institutions, a conservative estimate of distribution is shown below.

1. Breeding Station*

(FA Saladaga's Record)	Year	No. of Cuttings	Particulars
VSP 1-3	1983	134,350	Various Farmers
VSP 1-3	1984	940,630	Various Farmers
		126,900	DA (Regs. 7 & 8) Research Stations DCCLS (Tacloban) Walk-in request
		284,700	MAF Rehab Operating Program
VSP 1-4	1985	1,118,765	Farmers
		312,312	DA
		748,800	Philphos DCCLS LSBBA Walk-in requests MAF Rehab Operating Program
VSP 1-5	1986	193,547	Farmers
		30,000	MAF
		4,247	Walk-in requests
		34,140	Others
VSP 1-6	1987	103,050	Farmers
		30,000	DA
		161,200	LSBAA Philphos DCCLS Walk-in requests
VSP 1-6	1988	37,660	Walk-in requests including farmers experiment station
VSP 1-6	1989	53,900	Walk-in requests including farmers experiment station.
	Sub-total	4,314,201	

2. Training

VSP 1-6	1986-87	2,250	farmers/technicians
VSP 1-6	1988	8,120	farmers
VSP 1-6	1989	4,545	farmers
VSP 1-6	1990	940	farmers/cooperators
	sub-total	15,855	

3. Extension projects/Farmers' Field Day ViSCA Exhibits (PRCRTC Source)

VSP 1-6	before 1989	5,000	farmers
VSP 1-6	1989	4,060	farmers
VSP 1-7	1990	20,810	farmers
		700	BAMA/DCCLS
	sub-total	30,570	

4 Distribution by Non-NCT Cooperators (SCU's)

Catanduanes State College	(VSP 1-6, hybrids 3-190 15-70, 20-429)	32,000[1/]	farmers/DA
Central Mindanao University		50,000[2/]	farmers/other agencies
Northern Mindanao State Institute for Science and Technology		20,000[2/]	farmers/walk-in requests
Others		100,000[2/]	undetermined/some staff

5. Distribution by NCT

UP La Granja	30,000
CSSAC	210,000
Tiaong Experiment Station	37,090
DA, Bruane, Ilngao	600
Sub-total	277,690
Overall Total	4,840,316

---

* Includes all distributed taken from FA Saladaga and those distributed for ViSCA exhibits (Philtrade, Agro-fair, fiesta, etc.).

[1/] Does not include planting materials taken by farmers from experimental areas and those taken from the College of Agriculture not recorded by student assistants.

[2/] Conservative estimates

Data show that almost 90% of planting material distributed come from the breeding station and the highest agency recipient is the Department of Agriculture which in turn distributes to farmers or grow them in their respective propagation plots. Evidence seems to suggest a combination of relatively high mortality rate and non-adoption. Of the conservatively estimated 4.8 million vine cuttings distributed, at least 5000 hectares should have been seeded with the VSP's (i.e. assuming 33000 requirement per hectare and a cropping intensity of one). At present, less than one percent of total sweet potato area (about 1000 hectares) are seeded with the new varieties - mostly commercial and project/trial-linked areas.

Summary Characteristics of VSP Growers in Albay and Leyte, Cooperators and Non-Cooperator/adopters[1]

Variables		Cooperator	Non-Cooperator/Adopter
Years in farming general (yrs)		22	27
SP farming		22	23
Total farm size (hectares)		3.3	3.3
No. of parcels		3-4	2-3
Tenure:[2]	Owner-operator	60	51
	Part-owner/amortizing	-	11
	Tenant	60	75
Average total area planted to SP:			
(hectares)	VSP	0.7	0.3
	Local	0.7	0.4
SP % of total farm size		42%	21%
VSP % of total SP		50%	43%
VSP % of total farm size		21%	21%
VSP's tried:	1	80	43
	2	80	40
	3	80	9
	4	80	36
	5	60	15
	6	20	-
Still growing:	Y	60	38
	N	40	62
VSP Characteristics liked:			
early maturing		67	35
high-yielding		67	45
good eating quality		33	15
good market/acceptable to consumers		33	30
sweetness		33	75
Why not growing anymore?			
no more cuttings (severe rain)		50	21
no more cuttings (drought)		50	3
no market for wet SP		-	42
Source of knowledge: VSP's			
DA technician/extension		40	30
ViSCA project leaders		60	9
Landlord		-	4
Brgy, Capt.		20	-
Farmer-cooperator		-	24
Other farmers nearby		-	34
Source of planting materials			
SP buyer		-	2
ViSCA (requested)		60	4
Technician (ViSCA)		20	-
ViSCA project leaders		20	-
DA/MAO		-	19
Farmer-cooperator		-	43
Neighours/farmers nearby		-	30
Landlord		-	4
Cost of cuttings: free		40	70
sample		20	2
0.10/cutting		40	2
exchange		-	2
0.03-0.05		-	24

[1] Sample size, n, for cooperators is 5; n=53 for non-cooperators/adopters.
[2] Reflects parcels cultivated by a farmer have different tenurial arrangements.

IV. IMPLICATIONS

A study of the sweet potato seed production mechanisms is premised on the fact that it plays a critical role in the overall development of the sweet potato agro-industry. While such is recognized, there is little documented information on the various seed systems of sweet potato in the country. This study is an attempt to describe the indigenous production-distribution systems practiced by farmers, the official system established to evaluate and propagate improved planting materials and other alternative systems tried out to reach a greater number of beneficiaries. In addition, an attempt is made to assess the diffusion and adoption of the sweet potato HYV technology developed at ViSCA.

Findings show that only at best ten percent (10%) of sweet potato production has been linked to the formal seed system (i.e. HYV's); at least ninety percent (90%) thrive on the informal or indigenous farmer seed system. The reasons are: (1) the existence of a greater diversity of sweet potato across various agroecological zones in the country, each with an already existing farmer-selected good-performing cultivar; and (2) the lack of a systematic production-distribution scheme inherent in the weaknesses of a government-heavy set-up; and the lack of farmer-user participation and feedback. The bottom-up approach in varietal development and participatory method of seed production-distribution and monitoring lessen the risk of non-adoption.

The need to revise the official system to effectively develop and disseminate improved varieties had already been discussed in the early eighties triggered by low adoption of modern varieties and the dearth of quality seeds. The continued reliance, however, on the conventional, centralized, government-heavy formal system could bear heavily on scarce government resources, create a negative bias against farmer-user participation (thus, a higher risk of non-adoption), and a longer gestation for usefulness of the improved variety. The strictures imposed by the existing testing and evaluation protocol and of certification may not at all be relevant in a crop with a wide diversity like sweet potato. The evidence of unique adaptabilities of the crop to specific agroecological zones (e.g. upland, rainfed lowland, highland) renders the system of testing via the network's stations for national recommendation almost irrelevant and cost-inefficient. Even the stirrings for a regional recommendation seriously need caution. With the sweet potato farming circumstances, it is relevant to define region in terms of agroecological characteristics.

If the system of national recommendation is a means of recognition for achievement and national testing is then carried out, then a definition of desired characteristics for specific intended use and a description of testing zones are important for inclusion in the dissemination package, simply and clearly designed. In general, the system of evaluation for recommendation of sweet potato varieties could consider refinements in such aspects as: (1) criteria for recommendation qualified as to intended users (i.e. table vs. processing); (2) the basis for the selection of regional trial sites reflective of the typical sweet potato growing areas; and (3) choice of check variety or cultivar which should not miss the most widely grown or best performing cultivar of the test area. Performance failures of some high-yielding varieties in the stations are partly due to the relative lack of clear description of the recommendation domain.

The existing farmer seed systems have operated at a certain level of efficiency under existing scale of operations. These indigenous systems need to be considered in working out an innovative and simplified seed production-distribution scheme not excessively dependent on government but managed by farmers and supported by research and extension programs in the respective influence areas. With respect to this system, design may be area and use specific. A clear understanding of the size and nature of the seed demand, capabilities and strength of the farmer groups and local extension agencies, farming circumstances, and the fit of the improved variety to the market or use are critical inputs for an effective system.

Considering the enormity and nature of the tasks at hand, technical and social scientists need to collaborate and effect a medium by which a fuller synergy of constructive conflicts can be channelled - in the end, to help resource-poor farmers and rural entrepreneurs.

REFERENCES

Bartolini, P.U. "Storage, Germination, Establishment and Handling of Hybrids for Subsequent Evaluation and Dissemination of Sweet Potato" in Southeast Asian Germplasm Evaluation and Utilization. Nov 14 -Dec 12, 1980. ViSCA. Baybay, Leyte, Philippines.

Crissman, Charles C. Seed Potato Systems in the Philippines: A Case Study. International Potato Center (CIP). PCARRD Los Banos, Laguna, Philippines. 1989.

Feistritzer, Walther P. and A. Fenwick Kelly. Improved Seed Production. FAO. United Nations. Rome 1978.

Hansen, Esbern Friis. "Village-based seed production" in ILEIA Newsletter. December 1989. pp. 26-27.

Linnemann, Anita R. and Jau S. Siemonsma. "Variety Choice and Seed Supply by Smallholders" in ILEIA Newsletter. December 1989. pp. 22-23.

Monares, Anibal. "Analytical Framework for Design and Assessment of Potato Seed Programs in Developing Countries" in the Social Sciences at CIP. CIP. Lima, Peru. 1988. pp. 247-261.

Monares, Anibal. "Socioeconomic Research on Seed Production and Distribution System in Developing Countries" in Social Science Research and Training Report of Planning Conference. 1977. CIP. Lima, Peru. pp. 70-83.

Palomar, M.K. Sweet Potato Industry Report. 1989. Unpublished. Philippine Root Crop Research and Training Center. Visayas State College of Agriculture. Baybay, Leyte, Philippines.

Philippine Seed Board. Minutes of Meetings. Various Series 1969-1988. BPI-Manila.

Philippine Seed Board. Seed Catalogue. Bureau of Plant Industry. Manila. 1987.

Prain, Gordon and Urs Scheidegger. "User Friendly Seed Programs" in the Social Sciences at CIP. CIP. Lima, Peru. 1988. pp. 182-203.

Rasco, Eufemio T., Jr. "Plant Breeding, Seed Production and Distribution System in the Philippines" in Farmers Participation in Plant Breeding Research and Related Activities. Workshop. Sept. 13-14, 1990. UPLB. Laguna, Philippines.

Saladaga, Florencio A. "Sweet Potato Breeding in the Philippines" in Proceedings of the International Seminar on Root Crops in Southeast Asia: Production and Utilization. 1983. Manila, Philippines. pp. 371-378.

Schutjer, Wayne A. and Marlin G. van der Veen. Economic Constraints on Agricultural Technology Adoption in Developing Nations. Pennsylvania State University. March 1977.

IDRC SUPPORTED RAPESEED-MUSTARD RESEARCH PROJECT AT G.B. PANT UNIVERSITY OF AGRICULTURE AND TECHNOLOGY, PANTNAGAR, INDIA

Dr. Basudeo Singh
Intern Scientist (IDRC)
Department of Plant Breeding

G.B. Pant University of Agriculture and Technology,
Pantnagar - 263145, India

ABSTRACT

The case study pertains to a research project on rapeseed-mustard supported by International Development Research Centre, Canada at G.B. Pant University of Agriculture and Technology, Pantnagar, U.P., India from 1979 to 1986. The study highlights the importance of rapeseed-mustard in Indian oilseeds economy giving up-to-date data about area, production and productivity and its potential and place in present cropping sequences. India is the largest importer of edible oil. The major objectives under the project were development of high yielding varieties with better resistance/tolerance to major diseases and insect pests and better quality of oil and seed meal, and to conduct on-farm trials/demonstrations for the improvement and refinement of technology generated at the research station, and to transfer the technology generated under the project for adoption by the farmers.

During the operation of the project, 4 varieties of rapeseed-mustard were released and notified by Government of India. Their parentage, method of development superiority over standard variety/national check based on various yield and other tests in on-station, on-farm trials and demonstrations giving sufficient data have been described. The possible measured impact of the 4 new varieties in increasing the area, production and productivity of rapeseed-mustard in India have been presented by using official data. The systems of variety testing, release, notification, seed production and dissemination have been given in detail. Volume of seed production, the present system of on-farm storage, marketing problems and possible solutions have been discussed. Commercialization of the released varieties, their seed trade, Government policies and rights regarding cultivar release and control mechanisms have also been discussed in detail. The results of the case study have been summarised with conclusions drawn.

INTRODUCTION

Oilseeds, with an area of about 20 million hectares, claim the largest share in the country's sown area after food grains. These account for about 10 per cent of the Gross National Product (GNP). They are not only used for obtaining oils for cooking and other non-edible purposes but also as an input to many industries providing direct and indirect employment to millions of people mostly poor and landless. India was one of the oil and oilseeds exporting countries until 1961. Now it is the major vegetable oil importer. The reverse trend in the supply and demand was realised by the Government of India in the sixties. The Government of India launched an All India Coordinated Research Project on oilseeds from 1967 by establishing a number of research centres in different states and agro-climatic zones, and linked all the on-going ad-hoc or regular state Government-owned research centers with one another to avoid duplication and improve coordination. Despite this effort, the gap between supply and demand of vegetable oil kept increasing, leading to imports of vegetable oil beginning from 1975. Annual imports of vegetable oil currently average about US$555 million, though this is still increasing.

IDRC SUPPORTED OILSEEDS PROJECT

As already mentioned, the Government of India became very conscious about the supply and demand position of edible oil, and in 1976 approached the International Development Research Centre of Canada for its support in strengthening the on-going oilseeds research in India in selected centres. Consequently, IDRC supported oilseeds projects were launched at four centres: Pantnagar (Rapeseed), Hissar (Mustard), Jabalpur (Safflower) and Coimbatore (Sesame) from 1979 and continuing up to October, 1986. The present case study pertains to one of the above projects located at G.B. Pant University of Agriculture and Technology, Pantnagar, U.P., India. The project objectives were as follows:

Objectives

a) To breed high yielding, disease resistant, pest tolerant and widely adaptable varieties of rapeseed-mustard.

b) To screen germplasm for tolerance to high salinity, frost and freezing temperatures and drought.

c) To develop rapidly maturing varieties with high yield and high oil content.

d) To develop and adapt cultivars of good oil quality, low in erucic acid content which after oil extraction produce nutritious high protein meals low in nutritional inhibitors such as glucosinolates.

e) To develop suitable agronomic practices for the different agro-climatic zones.

f) To screen germplasm for better plant types well suited to intercropping of Brassica species with other food crops.

g) To conduct on-farm research and demonstrations.

PROJECT OUTPUT

The project concentrated on the development of varieties of rapeseed (Brassica campestris var. toria) and mustard (Brassica juncea) at the Crop Research Centre of G.B. Pant University of Agriculture and Technology, Pantnagar, Distt. Nainital, U.P. India. This Research Centre is located in the humid sub-tropical zone of the Himalayas at 29.0°N latitude, 79.3°E longitude at an altitude of 244 metres above sea level. The research team was led by the author as Senior Rapeseed Breeder since the beginning of the project till its termination.

Varieties Developed and Released

Four varieties (Table 1) of rapeseed-mustard were developed and released during the operation of the project. Most of the material was in the pipeline at different stages of development and testing with ICAR support at the time of inception of the IDRC project. The ICAR support continued with the IDRC support.

Table 1: Varieties of rapeseed-mustard released during project operation

Crop	Botanical Name	Variety developed/ released	Year of release
1. Mustard	Brassica juncea	1. Kranti	1983
		2. Krishna	1984
2. Toria	Brassica campestris var. toria	1. PT 303	1985
		2. PT 30	1985

The details of their parentage, testing and superiority are described below:

Mustard Varieties

1. Kranti - This variety was developed by pure line selection from a heterogenous population of Varuna variety, the best variety of mustard until 1983. It is a medium duration (125-130 days) variety with superiority of 14.7% over Varuna, based on a total of 19 location data from 5 years in the All India Coordinated Research Project on Oilseeds Trials, and from minikits in the farmer's fields. This variety is less susceptible to Alternaria blight and resistant to downy mildew and white rust. This was the first National variety of mustard released and notified for cultivation by the Central Seed Committee, Government of India in 1983 in all the major mustard growing states of India - M.P., U.P., Bihar, West Bengal, Rajasthan, Haryana, Punjab and Delhi.

2. Krishna - This variety was also developed by pure line selection from a heterogenous and homozygous population of Varuna, and released and notified by Government of India in 1984. This was the second National variety of mustard. Besides yielding 17.2% higher than the National check and parent population (Varuna), it has the widest adaptability as determined by its performance at 36 locations during 6 years in different agro-climatic zones. This variety is doing very well in Nepal also where it has been released by the HM Government of Nepal in 1988-89 for general cultivation in the same name.

Rapeseed (Toria) varieties

1. PT 303 - PT 303 is the first and the only national variety of toria. This variety out-yielded T9 (National check) by a margin of 12.5% in the All India Coordinated Research Project on Oilseed (AICORPO) Trials during 1976-77 to 1981-82 (6 years) over 20 locations, 16.6% cent in the U.P. State and university outstation trials during 1979-80 to 1981-82 (3 years) over 19 locations, and 9.7% in varietal demonstrations at the farmers' fields during 3 years (1981-82 to 1983-84) at 28 locations. This variety was developed from a cross between an early variety of toria - B 54, and a bold seeded variety of brown sarson - DS 17 MD, through recurrent selection. PT 303, besides being a higher yielder due to

more number of secondary branches, higher seed weight and more number of siliquae per plant, is tolerant to Alternaria blight, white rust and downy mildew, and gave 11.2% higher yield under diseased and unprotected conditions. It is one of the most widely adaptable varieties of rapeseed in India. This variety is doing very well in Nepal also where it has been released by the Government of Nepal in the name of 'Vikas' in Hindi, which means 'Progress' in English.

2. PT 30 - PT 30 variety of toria was released in 1985 by the State Variety Release Committee for cultivation in the sub-mountainous tract of U.P. called the Tarai region. Here it yielded 16% higher than the local and national check variety Type 9. It has better tolerance to Alternaria blight, downy mildew and white rust.

Varieties in Pipeline

At the time of termination of the project in late 1986, as many as 36 strains of toria, yellow sarson and mustard were under testing in various All India Coordinated Varietal trials and State Varietal trials. One of the toria types, PT 507, has been released by the Government of India in February 1990 for cultivation under rainfed situations in the Eastern States of India. One of the yellow sarson strains, PYS 842, has been recommended for release by the Annual Rabi Oilseeds Workshop 1990 for the state of U.P. Some are still continuing under testing.

Package of Cultivation Practices Developed

On the basis of agronomic, plant pathological and entomological investigations, a complete package of cultivation practices was developed for realising maximum yield, and this was recommended for adoption by farmers in the case of both toria and mustard.

On-farm and On-station Trials Demonstrations

Anticipating the release and notification by Government of India of a number of new varieties, varietal demonstrations and maximum production demonstrations on farmer's fields were started in 1980-81 using promising strains, which were later released as varieties, of toria and mustard. In the varietal trials, T-9, the standard variety, was used as check. The main objectives behind this advance action was:

i) To introduce the new varieties along with the standard variety to the farmers and show them at their own fields the superiority of the up-coming varieties.

ii) To disseminate the seed of new varieties to the farmers in advance.

iii) To educate the farmers about improved methods of cultivation, as field demonstrations are most effective tools among several means of communication to convince the farmer about the variety and the practices of cultivation.

Varietal Demonstrations

The results of varietal demonstrations of toria and mustard at the farmers fields are presented in Table 2 and 3 respectively.

Table 2: RESULTS OF TORIA VARIETAL ON-FARM DEMONSTRATIONS

Year	No. of demonstrations	Varieties/yield (kg/ha) T9	PT 30	PT 303
1980-81	2	1140	1127	-
1981-82	6	1392	1392	1391
1982-83	4	1156	1390	1363
1983-84	4	1206	1331	1325
1984-85	3	1242	1366	1392
1986-87	6	-	1181	1382
Overall mean		1191	1298	1371
Percent superiority over T 9 (standard check)		-	9.98	15.11

Table 3: RESULTS OF MUSTARD VARIETAL ON-FARM DEMONSTRATIONS

Year	No. of demonstrations	Varieties/yield (kg/ha) Varuna	Kranti	Krishna
1982-83	3	1350	1542	1543
1983-84	3	1475	1883	1650
1984-85	7		1603	1603
1986-87	6		1790	1768
Overall mean		1413	1630	1641
Percent superiority over check (Varuna)		-	15.35	16.13

The results of varietal demonstrations of toria (Table 2) revealed that on an average PT 30 and PT 303 were superior to check (T9) in yield by 9.98 and 15.1%, respectively, which was very close to their superiority of 12.51 to 16.6% in various on-station trials. This clearly indicated that for the farmers situation, the yield data of toria from station trials were as good as farmers field data because the varieties had stable performance and were definitely superior to the check.

Similarly, in the case of mustard varietal demonstrations on the farmer's field (Table 3) it was found that both Kranti and Krishna, newly developed varieties, gave 15.4 to 16.1% higher yield, and that their superiority in the farmers conditions was at par with that under station trials over the check variety, Varuna.

Maximum Production Demonstrations

In order to transfer the improved crop production practices and show the optimum yields of newly developed varieties, a series of maximum production on-farm demonstrations, each of 0.4 ha, were conducted with toria (Table 4) and mustard (Table 5).

In the case of toria, using PT 30 and PT 303 varieties (Table 4), it was observed that the average yields ranged from 1182 kg/ha to 1928 kg/ha, with an overall mean of 1364 kg/ha, which is almost double the national average yield. Similarly, the maximum production on-farm demonstrations of mustard (Table 5) conducted using Kranti and Krishna varieties revealed that under farmer's conditions, these new varieties yielded 1331 kg/ha to 1896 kg/ha with an overall average of 1715 kg/ha against the national average of 600 to 700 kg/ha, which means the productivity of mustard can be increased by 2 to 3 times using the new varieties and recommended package of practices.

Table 4: RESULTS OF TORIA ON-FARM MAXIMUM PRODUCTION DEMONSTRATIONS

Year	No. of demonstrations	Average yield (kg/ha)	Cost of production (Rs/ha)	Net return (Rs/ha)	Cost benefit ratio
1980-81	10	1261	1472	3573	1:2.43
1981-82	5	1389	1049	2837	1:2.70
1982-83	11	1182	1799	2932	1:1.62
1983-84	7	1374	1895	3600	1:1.90
1984-85	5	1370	1929	4236	1:2.20
1986-87	3	1192	2317	3641	1:1.57
1987-88	5	1829	2317	5914	1:2.25
Overall mean		1364	1940	3819	1:1.97

Table 5: RESULTS OF MUSTARD ON-FARM MAXIMUM PRODUCTION DEMONSTRATIONS

Year	No. of demonstrations	Average yield (kg/ha)	Average cost of cultivation (Rs/ha)	Net return (Rs/ha)	Cost benefit ratio
1980-81	3	1331	1509	3815	1:2.53
1982-83	2	1788	2467	4668	1:1.89
1983-84	4	1896	2487	4097	1:1.65
1984-85	6	1788	2650	5396	1:2.03
1985-86	8	1872	2858	4630	1:1.62
1986-87	6	1479	310	4381	1:1.41
Overall mean		1715	2628	4582	1:1.74

Using the Krishna variety of mustard, one maximum production demonstration of 0.4 ha was conducted in each of the five important mustard growing districts falling under the area of responsibility of the University. The results are presented in Table 6. The new variety yielded from 1500 to 2300 kg/ha under farmers fields, with an average of 1829 kg/ha, against the per hectare of mean productivity of 654 kg/ha. This clearly indicated that given the suitable practices and crop production technology, the new variety has the potential of increasing the yield by about three times the present average.

Large size mustard demonstrations of four to 10 hectares, called block demonstrations, were also conducted to verify the performance over large areas using newly released varieties in different districts during 1987-88. The results are presented in Table 7. In this case also, the results indicated that using new varieties and technology, mustard yield can be increased 2 to 3 times over the present yields.

Table 6: RESULTS OF MAXIMUM PRODUCTION ON-FARM DEMONSTRATION OF MUSTARD AND PRODUCTIVITY IN DIFFERENT DISTRICTS DURING 1987-88

District	Area (ha)	Demonstration yield (kg/ha)	Rapeseed-mustard productivity of the district (kg/ha)
Saharanpur	0.4	1500	661
Moradabad	0.4	2263	592
Badaun	0.4	1550	696
Bulandshahr	0.4	1531	661
Meerut	0.4	2300	661
Mean		1829	654

Table 7: YIELDS OBTAINED AT FARMERS FIELD UNDER BLOCK DEMONSTRATION AND PRODUCTIVITY OF THE MUSTARD IN THE CONCERNED DISTRICTS (1987-88)

District	Area (ha)	Demonstration yield (kg/ha)	Rapeseed-mustard productivity of the district (kg/ha)
Barielly	4	851	696
Rampur	22	2000	592
Moradabad	4	1850	592
Shahjahanpur	128	1600	602
Haldwani(Nainital)	4	1500	529
Rudrapur(Nainital)	8	1844	529
Mean		1608	590

Demonstration of Contribution of Individual Input Components

Research trials at the experiment stations are so well managed that all the inputs are provided to the crop up to the maximum economic level at most appropriate time. Thus the basic assumption in technology generation is that there are no resource constraints that the farmers face in crop production. However, the fact is that the majority of the farmers do not have enough resources to provide all the recommended inputs for maximum production even though they are aware of their importance. This results in poor adoption of technology. From the conduct of varietal and maximum production demonstrations, visits to the farmers fields and frequent interaction with the farmers, it was realised that the application of the full recommended package was beyond the capacity of the marginal and sub-marginal farmers who constitute the majority (74%) of the farming community. Therefore, experiments were laid-out to determine the contribution of individual components of the package of practices so that individual components could be listed in descending order for being applied by the farmers, depending upon their resources.

The study on the effect of important components on production in toria (Table 8) revealed that simple replacement of local variety by newly developed variety PT 303 resulted in an increase of 42.6 % in seed yield. The seed rate of toria being low (5 kg/ha), the improved variety seed was found to be the cheapest input with highest cost-benefit ratio, followed by new variety seed + fertilizer increasing the yield together by 56.5%. The adaption of full package increased the yield by 77.2. We advised the farmers that replacement of their local variety by the newly developed variety would bring maximum economic return per unit of expenditure. In the case of mustard also, the replacement of the old variety by the newly developed variety gave the highest cost-benefit ratio of 1:19, with a 20.3% increase in yield (Table 9).

Table 8: CONTRIBUTION OF IMPORTANT COMPONENTS OF PACKAGE OF PRACTICES IN TORIA (PT 303) CULTIVATION

Treatments	Seed yield (g/ha)	Increase in yield qtls	%
Local variety (Farmer's practice)	10.06	-	-
Standard variety (S.V.)	14.35	4.29	42.6
S.V. + Fertilizer	15.74	5.68	56.4
S.V. + Irrigation	13.52	3.46	34.4
S.V. + Fertilizer + Irrigation	17.17	7.11	70.1
Full package of practices	17.03	7.77	77.2

Table 9: CONTRIBUTION OF EACH PACKAGE OF PRODUCTION IN MUSTARD (KRANTI) BASED ON 5 YEARS AVERAGE

Treatments	Yield q/ha	Increase in yield q/ha	Increase in yield %	Cost benefit ratio
Local variety	5.8	-	-	-
Standard Variety (S.V.)	7.09	1.20	20.3	1:1.9
S.V. + Fertilizer	11.33	5.24	88.9	1:3.4
S.V. + Irrigation	8.86	2.17	36.8	1:5.4
S.V. + Plant Protection	10.53	4.64	78.8	1:8.3
S.V. + Fertilizer + Irrigation	14.43	8.54	145.0	1:4.5
S.V. + Irrigation + Plant Protection	10.19	4.30	73.0	1:4.1
S.V. + Fertilizer + Plant Protection	13.64	7.75	131.6	1:3.6
Full package of practices	17.78	11.89	201.8	1:4.7

On-farm Trials

Our work on on-farm trials was limited, and we were able simply to demonstrate, to the farmers, in their own farming conditions, the effect of individual components which are cheaper and more easily adaptable by the farmers, and which give maximum cost-benefit ratio. For example, in experiments at the experiment station it was found that sowing of toria in rows, 30 cm apart, enhanced yield by 15 to 25%. On-farm trials were laid out at farmers' fields to test the validity of the treatments at the farmers fields, it was found that line sowing was superior to broadcasting by 22% (Table 10).

Table 10: EFFECT OF METHOD OF SOWING ON SEED YIELD (KG/HA) OF TORIA UNDER ON-FARM TRIALS (1987-88)

Method of sowing	Locations/yield (kg/ha) 1	2	3	4	5	Overall mean yield (kg/ha)	Superiority (%)
Line sowing	900	725	700	770	800	779	22
Broadcasting	740	575	550	630	700	639	

As reported earlier in this section that experimental results clearly revealed that simple replacement of the local variety by newly developed varieties enhanced yield from 42.6% (Table 8) in case of toria and 20.3% in case of mustard (Table 9). Experiments were laid out at the farmers' fields to test the validity of the above findings under farming situation. The results are presented in Table 11. Improved varieties included in the trials were newly developed varieties of toria viz. PT 30 and PT 303.

Table 11: EFFECT OF VARIETY REPLACEMENT ON SEED YIELD OF TORIA UNDER ON-FARM TRIALS

Year	Variety	No. of locations	Mean yield (kg/ha)	Percent superiority over local variety
1984-85	PT 303	4	1773	17.8
	PT 30	4	1707	13.4
	Local	4	1505	
1987-88 (late sown)	PT 303	3	922	22.9
	Local	3	750	

The data presented in Table 11 clearly revealed that the simple replacement of local variety by newly developed variety enhanced the yield from 13.4 to 22.9%. This observation is similar to one obtained at experiment station. One of these trials was late sown, as some farmers are compelled to go for late sowing due to non-availability of vacant land on time. In this trial, other things being common, enhancement in yield was greater (22.9%) than under normal sown conditions. This may be attributed to the fact that late sown varieties are exposed to diseases and insects which appear late in the season and the newly developed variety, due to better tolerance to the pests, gives comparatively higher yields than local ones than under normal sown conditions.

Some on-farm trials were also conducted at the farmers fields to compare the performance of newly developed varieties of toria and mustard with the improved package of production, with local varieties with the local package of cultivation practices. The results are presented in Table 12.

Table 12: COMPARISON OF IMPROVED VARIETIES WITH IMPROVED METHOD OF CULTIVATION WITH LOCAL VARIETY AND LOCAL METHOD OF CULTIVATION IN TORIA AND MUSTARD UNDER ON-FARM TRIALS

Year	Crop	Variety	Cultivation practices	No. of locations	Mean yield (kg/ha)	Percent superiority over local variety + local practices
1988-89	Toria	PT 30	Improved	3	1187	24.7
		Local	Local	7	952	-
1989-90	Toria	PT 30	Improved	10	1499	48.3
		Local	Local	10	1011	-
1989-90	Mustard	Kranti	Improved	5	1020	28.3
		Krishna	Improved	5	1190	30.0
		Local	Local	10	915	-

The results presented in Table 12 show that increase in yield in case of toria with improved variety and improved package of practices over local variety with local package of practices ranged from 24.7 to 48.3% whereas in case of mustard this increase in yield ranged from 28.3 to 30.0%. Since replacement of local varieties with improved ones alone enhanced yields from 20.3 to 42.6% in mustard and toria, respectively, in other trials, the above increases with additional factor of improved practices are not significant. This is because of the fact that in the areas where these trials were conducted, local practices of cultivation were almost as good as improved practices. Therefore, most of the increases in yield could be attributed to change of varieties from local to newly developed ones.

During the conduct of on-farm trials and field demonstrations on the farmers fields, it was found that even in the best managed fields, the yields realised at research stations to determine the potential of the newly developed varieties were not repeatable. They reached 50 to 70% only of the yields obtained at research station. This was true even in the case of those farmer's fields which were located near the research station. However, in some cases crops and yields were as good as at the experiment station. These could be attributed to the variations in soil texture, structure, effects of previous cropping history on fertility, nutrient and micro-nutrient status of soil etc.

Impact of New Varieties on Area, Production and Productivity

The area, production and productivity of rapeseed-mustard since the release of new varieties is given in Table 13.

Table 13: AREA, PRODUCTION AND PRODUCTIVITY OF RAPESEED-MUSTARD AT NATIONAL LEVEL

Area - million hectare
Production - million tonnes
Productivity - kg/ha

Year	Area	Production	Productivity	Percent change in		
				Area	Production	Productivity
1983-84*	3.87	2.61	673	-	-	-
1984-85	3.99	3.07	771	3.1	17.6	14.6
1985-86	3.80	2.64	694	-1.8	1.1	3.1
1986-87	3.71	2.60	700	-4.1	-0.4	4.0
1987-88	4.50	3.31	748	16.2	26.8	11.1
1988-89	4.70	4.40	936	21.4	68.6	39.1
1989-90	4.8(p)	4.2(p)	875	24.0	60.9	30.0

* Base year as the first variety of mustard 'Kranti' was released in 1983 after crop season

p = provisional figures

A perusal of Table 13 would show that with the introduction of new varieties, there has been increasing trend in the area, production and productivity, with exception of 1985-86 and 1986-87 when a marginal decrease in area or production or both was noticed because of unfavourable weather and unfavourable market prices for the produce in the previous year. With the launching of the Technology Mission on Oilseeds in May, 1986 to give support price to the farmers and input on credit or credits for inputs and other necessary support and incentives, there has been significant progressive increase in area, production and productivity from 1987-88. Increase in area range from 16.3 to 24.0%, in production ranged from 26.8 to 68.6%, and in productivity ranged from 11.1 to 39.1% at the National level which is remarkable. The contribution of new varieties and technology is significant. Attempts have been made to quantify their contribution in the following pages.

Available Varieties

Forty two varieties of rapeseed-mustard have been released and notified by Government of India from 1961 to February 1990. This includes the two varieties of mustard and the two varieties of toria (rapeseed) developed during the Project leadership of the author duly supported by his team members.

Out of the above forty two varieties, there was demand for breeder's seed for just 18 varieties only; which means farmers have rejected 24 varieties at their own level. Since the farmer is the real judge of the suitability or otherwise of a variety for his cropping and climatic situation, the seed production plan is chalked out according to their choice of the variety and their relative seed requirement of each variety. Their anticipated requirement of certified seed is used in determination of the quantity of breeder seed required in case of each variety.

Estimated Area under New Varieties

It is rather a difficult task to determine the area under any variety of any crop in the absence of specific data. However, quantity of breeder or foundation or certified seed required and produced under each variety could give some estimate about the area under individual varieties. The sources of foundation and certified seed production are too many, and it is difficult to get the reliable data at national level from each source, but in case of breeder seed, all the seed requirements are pooled by the National Ministry of Agriculture, and all the breeder's seed required to be produced in case of any variety is done at one place only, usually at the variety-originating institution, under the supervision of the breeder who developed the variety or another qualified breeder. Thus the proportion of breeder seed required and produced in case of all the prevalent varieties could give an insight into the relative area covered by each variety. Actual area can be estimated if it is known what percentage of total area is under improved varieties. Therefore, the data available from Seed Division of the National Ministry of Agriculture, about the quantity of breeder's seed of each variety of rapeseed-mustard produced and allocated during 89-90 (Table 14) has been used in estimating the proportion and area under new varieties.

Table 14. BREEDER SEED OF PREVALENT VARIETIES OF RAPESEED-MUSTARD PRODUCED AND ALLOCATED (1989-90)

Sl. No.	Variety	Breeder's seed produced (qtl)	Sl. No.	Variety	Breeder's seed produced (qtl)
1.	Pusa bold	1.80	10.	Sangam	0.10
2.	Kranti*	1.25	11.	Bhawani	0.20
3.	Krishna*	0.40	12.	RLM-514	0.30
4.	RH-30	0.30	13.	T-9	1.36
5.	Varuna	3.15	14.	M-27	3.75
6.	RLM-619	1.12	15.	B-54	0.10
7.	TL-15	0.32	16.	Rohini	0.30
8.	PT-303*	0.42	17.	GSL-1	0.30
9.	RLM-198	0.24	18.	Vaibhav	0.05

Total seed of 18 prevalent varieties - 15.46
Total seed of new* varieties - 2.07

It would appear (Table 14) that out of a total breeder's seed of 15.46 qtls produced and distributed during 1989-90 in respect of 18 prevalent varieties of rapeseed and mustard, the share of new varieties developed under the project is 2.07 qtls, which works out to be 13.4% of the total breeder's seed. Thus it can be assumed that 13.4% of the total area under improved varieties of rapeseed-mustard in India is under new varieties. Official figures are not available about what percentage of total rapeseed-mustard area (4.8 million ha) is under improved and officially released varieties. However, unofficial sources claim that only about 50% (2.4 million ha) of this area is under officially released and notified varieties. On the basis of 13.4% area estimated above under 3 new varieties (one of the 4 varieties developed under the project, PT 30 is not in much demand as it has limited adaptability), actual area under these varieties can be estimated, which comes to 3.21 lakh hectare (13.4% of 2.4 million hectare).

Beneficiaries

C.S. Azad University of Agriculture and Technology, Kanpur, U.P. conducted a survey to find out the socio-economic factors associated with the cultivation of rapeseed-mustard and adaption of recommended package of practices in their cultivation. Among the various socio-economic, personal and psychological indicators of crops and technological adoption, the size of holding of farmers, risk factors, scientific orientation and crop productivity have been found to make significant impact. The survey revealed that generally the oilseed crops are grown by those farmers who have marginal and sub-marginal lands, and, based on smallholding size, are resource-poor, and unable to afford costly inputs like fertilizers, irrigation and plant protection chemicals. The survey also revealed that the poor farmers with low risk-bearing capacity mostly prefer rapeseed-mustard cultivation because of lower costs of production. It was also found that resource-poor farmers with low scientific orientation are more inclined to the cultivation of rapeseed-mustard, but the scientifically-oriented resource-rich farmers preferred wheat, which is an alternative crop of mustard despite the better cost benefit ratio in favour of mustard. Since edible oil is an essential part of human diet, almost every farmer with a low holding size of up to 2 ha. has no resources to buy edible oil from the market because of its high cost (US$2-3 per kg, depending on the brand). Therefore, in rapeseed-mustard growing areas of the country, most of the poor farmers grow rapeseed and/or mustard crop to meet their home needs, if only partially. Thus poor and small farmers are the major beneficiaries of the project.

Farmer's Criteria of Varietal Choice

Indian farmers have not had the appreciation for chemical quality characters like low erucic acid in rapeseed-mustard oil or low glucosinolate in seed cake. However, they have preferences for, and appreciation of, physical qualities of the produce like bold and shiny seeds and yellow seeds. The latter bring a premium price in the market because of the higher oil content and clearer oil from yellow seeds.

A survey to assess the criteria for the acceptance or rejection of a particular cultivar revealed that for their various cropping systems and crop rotations, criteria for varietal preference differ. In the multiple cropping system in which toria is taken as catch crop, the farmers prefer early maturity varieties for two reasons: (i) an early variety will vacate the land in time for the sowing of a following crop, such as wheat (delayed sowing of wheat, reduces yield significantly; thus the farmer is not ready to accept a reduction in wheat yield in favour of higher yields of toria coming from late maturing varieties), and (ii) early maturing varieties, because of early harvest, escape from damage caused by diseases and insects which appear late in the season. This eliminates the use of plant protection chemicals.

If a farmer is to grow sugarcane, a fodder crop, spring season vegetables or pulses, which are planted in the month of February/March, he prefers to grow late maturing varieties, which invariably give higher yields than early maturing ones. However, a farmer would prefer only those late varieties which have resistance or tolerance to prevalent diseases and insects. Similar, with mustard, varietal preference of a farmer depends upon the crop rotation he is to follow. In the case of a short period between two other crops in the rotation, a farmer prefers early varieties even with low yields, but where the period is longer, he would grow late-maturing, higher-yielding varieties. For other specific situations, varietal preferences differ according to the location of land and its soil type. The rainfed farmers require drought resistant or tolerant varieties, whereas for alkali soils, the farmers' preference is for salt-tolerant varieties.

CULTIVAR TESTING, RELEASE AND NOTIFICATION SYSTEM

Each variety under development is subjected to early generation testing at the Experiment Station by the breeder. This is based on the assumption that populations or lines with better potential can be identified on the basis of their performance in early tests. From the varietal evaluation trial usually called 'Station Trials', on the basis of their consistently better performance for 2-3 years, the one to three best materials are identified for trials in different agro-climatic zones through the All India Coordinated Research Project on Oilseed (AICORPO) Trials.

Varietal Testing and Release at National Level

For the purpose of AICORPO Trials of rapeseed-mustard, data of which are considered for release of the variety at national level, the rapeseed-mustard growing areas in the country are divided into 3 zones: (i) North West Plains Zone (NWPZ), (ii) Central Zone (CZ) and (iii) Eastern Zone (EZ). These zones have been delineated on the basis of agro-climatic conditions as well as cropping systems. In an annual oilcrops workshop, every breeder proposes 2-3 varieties of rapeseed-mustard, supported by Station Trial data about its potential and likely superiority, for inclusion in the first stage of AICORPO Trials (Initial Evaluation Trial, IET), followed by evaluation in Coordinated Varietal Trial (CVT), National Evaluation Trial (NET) and Minikit if found superior to three checks at each stage of Testing. The three checks are (i) the national check, (ii) the zonal check and (iii) the local check. Each of the trials (IET, CVT and NET) is conducted in successive years.

When data become available from IET, CVT, NET and Minikit, the breeder submits the release proposal before the Central Sub-Committee on Release and Notification of Varieties. After approval from this Committee, the variety is released and notified for the cultivation and production of seed.

Constraints and their Solutions

1. Until 1982, no variety could be proposed for release unless it had been tested for at least six years in various coordinated trials (IET, CVT and NET). This was an unusually long time for variety testing. Since 1982, the number of testing sites has been increased, and six years of testing have been reduced to three, in order to release the superior varieties more quickly.

2. Minikit data from the states for which a variety was likely to be released was obligatory for submission and consideration of release proposal until 1985. However, it was found that the Department of Agriculture generally did not evaluate the minikits, or did not supply the data, or, if the minikit was conducted and the data supplied, that the data were unreliable. This resulted in the withholding of some release proposals for 2-3 years. In order to overcome this problem, it was decided that minikit should be evaluated at least at 10 locations in farmers' fields by the cooperating breeders, and reported at the workshop. These data would be considered sufficient for the purpose of release of the varieties in place of minikit data from the states.

3. In order to obtain farmers' field data and farmers' reaction before a variety is proposed for release, the author invariably goes to the farmers with the new varieties still under test, if they appear to be very promising in the first year of AICORPO trials. Varietal demonstrations are conducted, using the best released variety(ies) as check(s). Thus by the time the variety has been tested under AICORPO Trials and identified for release, at least two years farmer field data and farmer reaction are already at hand, which are of considerable help in variety release. This system could be adopted to any breeder's advantage.

System of Testing and Variety Release at State Level

The State of U.P. has 8 agro-climatic zones, and there is at least one Agricultural Testing and Demonstration Station in each zone. The main function of these stations is to conduct State Varietal Trials, testing the suitability of new varieties contributed by crop breeders of State Agricultural Universities, Central Universities or even private seed companies. There is a State Variety Release

Committee with the Director of Agriculture of the State as its Chairman. Additional Directors and Joint Directors of Agriculture of the State, Heads of Department of Plant Breeding, and Directors of Research of three State Agricultural Universities, and officers-in-charge of all the 8 regional agricultural testing and demonstration Stations are members of the State Variety Release Committee. This committee performs two functions: (i) planning of the State Varietal Trials and (ii) release of the varieties if found suitable after 3 years of testing in State Varietal Trials.

SEED PRODUCTION AND DISSEMINATION MECHANISM

For better understanding of seed production and dissemination, it is necessary first to know the kinds of seeds produced, the agencies or organisations which are involved in seed production and dissemination, and the latter's function.

Kinds of seeds

1. Breeder's Seed

The seed produced directly under the supervision of the originating/sponsoring breeder or institution is called Breeder's Seed (BS). It has the maximum genetic purity and is the source of production of other classes of seed.

2. Foundation Seed

The progeny of breeder seed used to maintain specific genetic purity and identity is called foundation seed. The production must be acceptable to certifying agency. It is the primary source of seed of a genetically-identified variety from which all further increases are made.

3. Certified seed

The progeny of foundation seed so handled and produced as to maintain satisfactory genetic purity specified for the crop is called certified seed. The production must be acceptable to a certifying agency. This seed is given to the farmers.

In some crops like groundnut where the seed rate is high and seed multiplication ratio is low, one more multiplication of foundation seed as well as of certified seed is also done to produce foundation stage II seed and certified seed stage II seed. However, in case of rapeseed-mustard, this is not required because of low seed rate and high seed multiplication ratio.

Organisation of Seed Production Agencies

There is a National Seed Corporation (NSC) and a number of State Seeds Corporations which are responsible for organising the seed production program. Almost every state has its own seed corporation. However, there are still some states which do not have seed corporations. In such states, the function of seed corporations is performed by the NSC. The state seed corporations, jointly owned by state government, NSC, seed growers and state agricultural universities, have their major function of production, procurement, processing, storage and distribution of required quantities of seeds of each crop and variety as assessed and/or required primarily by the concerned state Department of Agriculture and Government of India. These corporations work usually on commission on a no loss, no profit basis. However, these corporations have no authority for seed certification, this being done by another independent state agency, the State Seed Certification Agency (SSCA).

Organisation of Seed Certification Agencies

Like state Seed Corporations, the State Seed Certification Agency (SSCA) set up in almost every state is an autonomous body managed by a Board of Directors which includes representatives of various interests, namely, Department of Agriculture, State Agricultural Universities, Specialists (e.g. breeder, pathologist, entomologist), seed producers, and representatives of Seed Certification Agencies and Seed Analysts.

Seed Distribution Mechanisms

There is a number of mechanisms in operation for distribution and dissemination of the seed of the improved and new varieties. The dissemination of seed of the varieties was started even when these were in advance stage of testing after it was realised through the results of trials at the research stations, AICORPO trials and varietal trials at the farmers fields, that they were best and likely to be released. The minikits were prepared in thousands every year to supply farmers visiting the university during farmer's fairs, the lab-to-land program, and for training. Seed dissemination also took place simultaneously through farmer-to-farmer sale from the harvest of on-farm trials and demonstrations. G.B. Pant University grows more breeders seed than is required by the Government indent system, and has sold this seed to farmers as Truthful-Label-Seed more than two years ahead of official release of the variety (as much as 700qq in one year). Thus the varieties had alreadly spread far and wide and become very popular much before their formal release and notification. Other ways of seed dissemination started after their release and notification. These are listed and described below.

1. Government Seed Stores

State Governments have opened a number of seed stores and one seed store is responsible for the supply of certified seeds and other inputs to 30 to 40 villages. In areas where there is less coverage, Government seed stores are supplemented by consumers' cooperative stores.

2. Seed Corporation's Distributors and Retailers

Each state seed corporation and national seed corporation have appointed their seed distributors and retailers in the area of their operation. These distributors and retailers also sell the seed to the farmers.

3. Government, Minikit Program

The central as well as state governments have seed minikit programs, in which 1-2 kg seed packets along with the printed literature on crop production technology of the crop and necessary fertilizers and pesticides sufficient for area to be covered through this seed, are supplied. For this purpose, the potential areas and locations are identified at the national and state level. This is done to popularise the best varieties as well as motivate the farmers to increase the area under oilseeds. This is being done by the MOA through a special project called NODP (National Oilseeds Development Project) with an outlay of INR 1700 million (US$100 million). This is to be shared equally between central and the state governments, to step up oilseeds production in 18 selected districts in 17 states. Another special project, Oilseeds Production Thrust Project (OPTP) with an outlay of INR 1250 million, equivalent to about US$74 million, is also under operation in 246 districts in 17 states including 151 NDDP districts.

4. On-farm Demonstrations

On-farm demonstrations are conducted by State Governments under centrally and state sponsored oilseeds development and extension programs. On-farm demonstrations are also conducted by the oilseeds project scientists of agricultural universities and institutes in the areas which fall within the jurisdiction of their centre/institution. Farmers are supplied the seeds of improved varieties and other inputs. Thus the farmers receive the pure seeds of improved varieties.

5. Oilseeds Growers Cooperative Federations

The National Dairy Development Board (NDBS) initiated oilseed growers cooperative federations in seven states, covering 13 lakh hectares under oilseeds with an investment of INR 2500 million, equivalent to US$147 million. Plans for additional coverage of 25 lakh hectares involving INR 4500 million, equivalent to US$265 million, are underway. These cooperatives, besides procurement and supply of quality seeds of improved varieties, also provide other input support on credit to the farmers, and arrange for procurement, storage and marketing of the produce. They have even set up processing and packing industries for oil extraction and its sale in small disposable containers for the benefit of consumers.

6. Seed Corporations

National seed corporation and state seed corporations also supply the quality seed of improved varieties directly to the farmers, in addition to their normal channels of distributor and retailer through their own outlet and sale points.

7. Private Seed Companies

A number of private seed companies are also engaged in the production, processing and sale of quality seeds of varieties developed by the companies, as well as by public institutions. They have their own trade name. They generally do not go for certification. However, more often than not, the seed supplied by the private seed companies are superior to the certified seed of the same variety supplied by Government seed corporations. They maintain seed quality and a more efficient supply system to maintain their position in the seed market.

8. Agriculture Allied Industries

Public and private-sector agriculture-allied industries engaged in the production of agro-inputs such as fertilizer and pesticides, spend a part of their turn-over on rural development. This activity also involves supply of quality seeds of priority crops and other inputs, and conducting on-farm demonstrations on varieties and crop production technology. Sometimes, they get the demonstration plots registered for production of certified seed, which is used next year for free distribution among the farmers.

9. Farmer's Fairs and Meetings

Almost all the agricultural universities and research institutes organise farmers' fairs at their main campus and research stations. These also organise farmers contact program as well as village and divisional level meetings. On all these occasions, farmers are provided 1-2 kg seed of latest improved varieties, which they grow themselves and distribute their seed through sale to their fellow village farmers and their relatives. Thus the seed of new variety gets disseminated to a large area.

10. Oilseed User Companies

Oilseed processors and exporters have their own obvious interest in the abundant supply of raw material. All such companies spend some money to promote oilseed development in the area of their operation. In this direction, their first step is to conduct demonstrations with latest varieties on the farmer's fields and supply them quality seed of best available varieties. Thus they also contribute to the varietal dissemination.

11. Farmer to farmer transfer

Farmer to farmer transfer of seed is very common. If a farmer obtains the seed of improved varieties either through minikit program or by any other sources and grows the same in his field, the neighbour farmers seeing its better performance get motivated to grow the new variety. They may receive the seed from their neighbour farmers, though sometimes they have to pay a much higher price.

Constraints in Seed Distribution and Possible Solution

1. Rapeseed-mustard seed because of high oil content loses seed viability quite fast under ordinary storage conditions if it is not packed in moisture-proof containers. Seeds should be packed in moisture-proof containers, so that not only deterioration in germination can be prevented but also it is made suitable for carry over for next year.

2. Majority of the oilseed farmers (74%) have their land holding size below two ha, in which they grow other crops also. The seed rate being low (5 kg/ha), their seed requirement is small but the certified seeds are packed in 5 kg and 10 kg cloth bags. Therefore, small farmers do not require one full bag of seed. This forces them to locate other needy farmers and pool their requirement equivalent to the seed in a bag and then purchase a bag and distribute among themselves. To overcome this problem, it is necessary that seeds should be packed in 1/2 to 1 kg bags more than in the larger bags.

COMMERCIALIZATION AND RETURN

Rapeseed and mustard are very important amongst 7 edible oilcrops grown in India and account for about 28% of the total edible oilseeds production, next only to groundnut. But in terms of potential for further growth, these are most important and therefore figure prominently in the National Oilseeds Development Program of Government of India. These crops are already very important commercially. Rapeseed mustard are grown as edible oilseed crops as well as cash crops in 18 of the 21 states in India.

Involvement of Traders and Producers

Over 4 million tonnes of rapeseed-mustard are produced annually. A production of this magnitude naturally involves millions of people in production, transportation, storage, processing, packing and marketing. Traders are very active in rapeseed-mustard marketing, because the seeds of these crops can be stored for a long time with a near absence of any problem from storage pests. Since these are grown as cash crops also and are harvested at such time when no other crop is ready for harvest (this is true particularly in case of toria which is grown mostly as a catch crop), and farmer needs cash for the following input-intensive staple food crop - wheat, the market is flooded with seeds of these crops, and in view of the abundant supply, traders force the farmers to sell at lower prices. Traders do maximum purchases and store the seed and create artificial scarcity in the market to raise the sale price to make maximum profit in the lean season.

In order to save the farmers from the exploitation by the traders, the Government announces the minimum support price before harvest. This means that if the ruling price in the market goes below a particular level, the Government will come into the market to purchase the seed from the farmers. But it has been seen that ruling prices are mostly higher than Government support price. Thus the minimum support price has hardly helped the farmers in rapeseed-mustard. The Government fixes the support price on the basis of cost of cultivation with very little margin of profit.

Reward for the Project for Variety Development and Seed Supply

Development of varieties, cultivation and plant protection practices are normal functions of any crop improvement project in India. Project personnel are regular employees of the agricultural universities and research institutions on full time basis on monthly salaries, and with other facilities. Project output belongs to the organisation, not the project scientists. Thus there is no system of direct reward to the project staff. However, project staff can claim credit for the work whenever they are assessed for higher positions within or outside the organisation.

In the present case, when the IDRC project was sanctioned by the Indian Council of Agricultural Research (ICAR) for implementation at Pantnagar in 1979 when Pantnagar was a testing centre for rapeseed-mustard research, the IDRC project brought more staff and facilities and the Centre was strengthened to a great extent. By the time the IDRC support came to an end in late 1986, the Centre had already contributed significantly to the research efforts of the State and Central Government through ICAR in the form of 4 varieties already released with many more in the pipe-line at various stages of testing in the AICORPO trials at the National level, and in State Varietal Trials at the State level. Its contribution was realised at National level. The status of Pantnagar was raised to that of Centre of Excellence for rapeseed-mustard research in the country, and ICAR, after withdrawal of IDRC support, not only increased its support to the IDRC-level on a regular basis, but also provided more staff and funds for future work. In the subsequent IDRC-supported project on rapeseed-mustard for inter-institutional collaboration between some Indian and Canadian Institutions, Pantnagar Centre has been given the role of a main Indian Centre with 4 of the 5 mandates in the collaborative project assigned to it. The project personnel consider this recognition a big reward, and appreciation and recognition of their work.

Once a variety is released and notified, the responsibility of the organisation and the breeder developing the variety is to continue to produce nucleus and breeder seed. The requirements of the breeder's seed from all the states and organisations are compiled in the Seed Division of the Ministry of Agriculture, Government of India, which passes them on to ICAR, which in turn passes them on to the concerned breeder/ institution. In case of oilseeds breeder seed, 50 percent of the anticipated cost of breeder seed indented by ICAR is given to the institution as an advance along with the indent, and the actual balance amount is paid on the basis of the actual quantity of the seed produced and supplied to the designated agencies.

Commercial Seed Production

Commercial seed production has become very popular particularly among the big farmers in the recent past due to higher margin of gains per unit of area and production. About 12,000 to 16,000 quintals of breeder, foundation and certified seed of the 4 varieties of toria and mustard developed under the IDRC-assisted project are produced every year. At current prices, the volume of this seed trade is of the order of INR 18-24 million (US$1-1.4 million) per annum.

VIII. SUMMARY AND CONCLUSIONS

1. In India, oilcrops, occupying about 20% of the total cropped area, claim largest share in country's sown area after food grains, and account for 10% of the value of all agricultural products and 5% of the Gross National Product. The per caput availability of edible oil at 5.9 kg/adult/year (1987) is far below the critical level of 13.14 kg, as well as the world average of 12.23 kg.

2. During the operation of this project, two varieties of early maturing toria (*Brassica campestris* var. toria) namely, PT 30 and PT 303, and two varieties of mustard (*Brassica juncea*) - Kranti and Krishna, have been released and notified by the Government of India for general cultivation in the country. PT 303 was the first national variety of toria released in 1985, and Kranti and Krishna were and first and second national varieties of mustard released and notified in 1983 and 1984, respectively. PT 30 was released as a state variety 1985. These varieties are higher yielding by 12-18% over the best standard check and have better tolerance/resistance against major diseases and insects. These varieties are being grown quite extensively and annually occupy an estimated area of 3.20 lakh ha., which forms 13.4% of the total rapeseed-mustard in India under improved varieties. The annual breeder, foundation and certified seed sale is of the order of 1 to 1.4 million U.S. dollars. Two of the above varieties have done well in Nepal, and have been released by that Government for cultivation.

3. Considering area, production and productivity of 1983-84 as base data, it has been calculated that the increase in area under oilseeds ranged from 16.3 to 24.0 %, in production ranged from 26.8 to 68.6 percent and in productivity ranged from 11.0 to 39.0 %. Considerable contribution to this increase, particularly during the last three years, can be attributed to these new varieties. They occupy about 13.4 % of the total rapeseed-mustard area under improved varieties which comes to 3.21 lakh hectare.

4. Major beneficiaries of the new varieties and technology are marginal, small, resource poor and less educated or uneducated farmers who constitute the majority of the rapeseed-mustard growing farming community.

SEED PRODUCTION AND DISSEMINATION MECHANISMS WORKSHOP

CASE STUDY REVIEW

Neil Thomas, Study Coordinator

This workshop has focused on the experiences of seven IDRC projects active in the general area of plant breeding or technology generation and transfer. The presentations have shown us that these projects have been active in multiplying and disseminating improved materials produced by the projects, and that the projects have had close links with the rural communities for which these materials were intended. What characteristics of these projects are of particular interest to our pursuit of effective mechanisms? Can we draw any specific lessons?

The Vegetable Seed Project in Northern Thailand shows that it is feasible to produce seed of sweet corn and crucifers in this area. It shows that some of these may be effective farm-level substitutes for the opium poppy in terms of income-generation, though there are still some marketing constraints to solve. The Project is well-placed to feed its own varieties into this production system. The Project has elucidated differences among the Hill tribes in their interest and capability in technology adoption, and has suggested ways of capitalizing on these differences. Having worked so closely with small farmers, the Project has learned which crops are of interest to the farmers, and how these people prefer to grow them. In the last year it has seen a significant increase in spontaneous adoption of one seed crop, Chinese radish. Of importance is that the Project developed seed production systems prior to the release of its own varieties, i.e., it has already developed on-farm systems that the farmers have adopted. However, the marketing of farmer-produced seeds is still a concern, and will be crucial to seed production being a sustainable alternative in these cropping systems. To date the Project has bought back all farm-produced seeds.

The Pigeon Pea Project in Kenya has been successful in achieving dissemination of its new varieties. Several mechanisms were tried, including an existing integrated rural development project which was looking for appropriate interventions. The Project itself is still the largest multiplier and provider of seed, due principally to little interest on the part of the private sector in a rainfed crop for marginal areas. Successful adoption was due to a variety which was significantly shorter-seasoned than, though with similar quality characteristics to, the traditional variety. The Project will try private-sector marketing in the coming year.

The Rapeseed Project in India has released five varieties which now occupy a significant proportion of the total area of the country planted to oilseed crops. The material has also been adopted in Nepal through free trade across the border. Throughout their development, these varieties were tested in on-farm trials, and as a result of large releases of seed by the originating institution, were seeded over a significant area well before their official release. Minikit testing was also carried out, as is common to all crop varieties in India. Undoubtedly important to widespread adoption was the superiority of these materials when compared to previous varieties, but the unique ability of the originating institution to move breeder's seed directly into widespread multiplication cannot be overlooked in examining the rate of adoption.

The Sweet Potato Project in the Philippines has had material disseminated throughout the country. Some adoption was due to the rapidity which earlier varieties provided a crop, an important characteristic after decimation of other crops by typhoons. To increase adoption rates, which appear to have been limited so far, the Project team has focused on the different quality characteristics of each cultivar, developing specific industrial processes for them. It is not known how the availability of this technology has influenced adoption. Later varieties have been closer to traditional varieties in their qualitative characters. Informal channels have been far more important factors in dissemination than formal ones, suggesting that efforts to create a more effective formal system may not be worth the cost.

The Andean Farming Systems Project in Peru has established revolving funds as a way of disseminating new varieties of various crops throughout campesino communities in the Andean highlands. Originally started as a service component to a broader research program, this component is now involved in disseminating other aspects of the production technologies of these crops. Potato is the principal crop being dissmeminated. The marginal environment of this region has shown the fragility of crop production, and thus of any mechanism designed to be continuous across years, especially under campesino management. Clear from this project is the need to focus as much research emphasis on dissemination and campesino adoption as on the biological processes of varietal improvement. These funds were established at a time when Peru was passing through a period of hyperinflation, and the effects of this have been evident. It may be questioned whether government agencies have the entrepreneurial skills to operate these funds successfully, both administratively and financially. At this time, these funds have not shown themselves to be sustainable. They require a high degree of managerial capability.

Two other Projects included in the case studies, but not represented at the workshop, have both achieved some success in dissemination. This is clearer in the Food Legume Project in Pakistan, where new varieties, especially of chickpea, have found considerable acceptance. It is less clear what acceptance the millets of the Millets Project in India have had; the author reports that the area seeded to millets is under a continual decline even though new varieties are being developed. Millets, particularly, are a crop of marginal areas, where few possible alternatives exist for the small farmers of such regions. The manuscript of the Food Legume Project only is included here.

General lessons learned

1. These projects have been flexible in their multiplication and dissemination of material. Farmers have been given access to the material during its development, and have been involved in on-farm testing. This has resulted in high rates of adoption in relatively short periods of time. In most cases, seed has been available when farmers have wanted it.

2. Where fairly rigorous state-controlled systems of varietal screening and release exist, they have neither contributed to nor constrained rapid dissemination, having perhaps operated more as official baptizers of material already widespread. Informal dissemination systems have been the most significant channels in adoption, and local organizations may have been significant contributors.

3. In general, production issues do not seem to be as critical in dissemination and adoption as marketing issues. Projects generally have not considered

downstream aspects, or have not been able to establish sustainable marketing mechanisms. In the traditional sense, it might appear to be too early to consider these, though an holistic approach to crop improvement would suggest that such issues should be built in from the start.

4. Projects dealing with marginal crops have difficulty in assuring private-sector involvement in seed supply. Demand exists generally only in years after on-farm production shortfalls, when farmers cannot save enough of their own material. Continued public-sector involvement in one form or another may be necessary to ensure continued availability of improved varieties.

5. Revolving funds do not necessarily guarantee continuity in the resources necessary to ensure availability of planting materials. Revolving funds require a high level of managerial capability, which is not often available in public-sector institutions or farm communities.

6. Had these case studies not been undertaken, it is questionable whether most projects would have been sure of the amounts of improved material reaching farmers, and thus the potential benefit of these varieties. At the outset, projects' staff were sensed to be ambivalent about the usefulness of such studies. The results show, that, in some cases, the actual amounts moving through different channels still have not been well quantified. In these projects, as in many others, often there is not a need felt by staff to know precisely how much is moving, as long as there is evidence that some is. This is a typical public-sector response, and does not argue for efficient use of resources in achieving objectives.

7. Projects need more clearly defined methods to allow measurement of the amounts of seed moving through different channels.

Guidelines for the future

These experiences, and others from projects elsewhere, should be used in improving the efficiency and effectiveness of crop and varietal improvement projects, and the small-farmer-oriented seed industry as a whole. Perhaps the most important issue is that of the involvement of the farmer in the improvement or development process. If this is a basic precept, then the following suggestions can be made:

1. View plant breeding in a wider context. It should not just be a biological process conducted in an institutional environment, rather it should be an interactive multidisciplinary process where the end-user is involved in the definition of the improvements intended, and the evaluation of the material as it is produced. This may even extend, as has been suggested before, to the farmers operating low-cost trials networks.

2. A project should document the process of on-farm testing and development much more completely than is currently done. This means looking beyond the simple measure of improvement in the prime characteristic under study, to recording the farmer's management, and what he or she does with the material once the test is over. Where possible, evaluative procedures should be built into these activities, to determine whether, both in the farmer's and the researcher's eyes, worthwhile gains are being made.

3. Encourage the public-sector institutions which view their function as one of regulation to look more towards the provision of technical assistance and

training to the small-scale seed industry. Attempted regulation of this sub-sector, beyond the provision of basic guidelines which training would encourage be accepted, will stifle entrepreneurial spirit.

4. Where a donor agency, such as IDRC, commonly funds several projects in a given sub-sector, a generic framework for definitive issues and data collection would be extremely useful. Many of the IDRC projects reviewed here were developed independently, and, while successful in meeting their specific objectives, pose difficulties when evaluation (especially comparative) is considered. Such projects should collect a minimum set of baseline data for future impact evaluation. This dataset should be specified in the generic framework. A minimum data set should cover (it should be noted that some issues may be qualitative):

    i) Characteristics of target group(s), including farming system(s).

    ii) Production and its constraints in crop being improved, including average yields, losses due to pests and diseases, agroecology.

    iii) Marketing and its constraints in crop being improved, including on and off-farm flows, costs and returns, demand and stability.

    iv) Sociocultural issues which impinge on the farming system, and which may affect labour availability, acceptance of new varieties, entrepreneurial capability. Importance of local organizations in the farming community, and what they do.

    v) Existing extension systems, and methods used in reaching farming community.

    vi) In relation to iii), the dissemination mechanisms common to the target group, and the proportions of material that flow through each. An analysis of the relative efficiency of each.

During the course of a project, the staff should review these aspects on a regular basis, updating their baseline information. Rapid rural appraisal could be a particularly useful technique for this purpose. Particular attention should be paid each year to the changes in the target crop, especially in farmers' responses to materials being tested on-farm. Information of the type listed here is necessary for ex-ante analyses which will justify the effort to be dedicated to a breeding program.

GENERAL GUIDELINES FOR THE DEVELOPMENT OF SEED SUPPLY UNDER MEDIUM AND SMALL FARMERS SITUATIONS

Adriel Garay

Based on the experience of all participants, some principles were identified as being useful in the process of developing effective supply and adoption of improved seeds. This exercise was carried out concentrating on four key topics:

1. Institutional organization of the seed supply system:

   - Make sure that the institution that generates the variety produces and supplies sufficient quantities of basic seed on a timely and continuous basis, to those organizations interested in commercial seed production.

   - Avoid conflicts of interest by organizing production and marketing of seeds with organizations that are not involved in official seed control service.

   - Develop simple and functional mechanisms avoiding complex procedures and regulations.

   - Promote and capitalize on the participation of other organizations in the production and marketing process. These organizations may include development projects, universities, NGOs, farmer-producer organizations, available in the target region or country.

2. Seed Production Process, Methods and Tools

   - Promote local (grassroots) production/marketing groups within those regions where demand is anticipated and provide assistance to train them in seed production and marketing.

   - Introduce scientifically sound seed management methods to assure and maintain quality in the post harvest phase of production.

   - Incorporate simple but effective methods of product differentiation to facilitate the identification of improved seeds, and gradually to educate the consumer on the benefits of using improved seeds.

3. Marketing

   It was indicated that most projects were production-oriented and do not pay enough attention to market assessment, and marketing processes. Some of the key features in marketing were identified:

   - Timeliness in seed delivery. Make seeds available when needed.

   - Make seeds available where needed. This is particularly necessary when supplying seeds in regions with limited transportation facilities. Avoid long travelling distances.

   - Differentiate the improved seeds by using clearly printed labels or bags, so that the potential buyer can differentiate from common grain.

- Let the free market rules of supply and demand work. Do not artificially control prices.

4. Training:

   - Train farmers in methods that contribute to high yields of quality seed at low cost.

   - Train technicians to provide the technical support to seed producing farmers/organizations.

   - Train in production and marketing methods and strategies.

   - Utilize the "learning by doing" or training in action method as a follow-up to classroom training.

During the discussions, two points of strategic nature were also identified:

- Excellence in biological research is not enough, especially when addressing medium and small farmers, to achieve transfer and adoption of improved technology. Decided and clearly focussed actions, such as development of seed supply mechanisms, are needed.

- In developing seed supply under medium and small farmer conditions, flexible and small grassroots approaches seem to facilitate the process better than trying to conform to existing rigid structures when initiating the process.

REVIEW OF MEETING'S GOALS AND ACHIEVEMENTS

Nicolas Mateo, Associate Director (Crop Production Systems)

Departure points for IDRC:

Where have 20 years of Centre support to plant breeding led to? What are the positive and negative lessons learned? Should IDRC continue supporting plant breeding efforts at all? If yes what strategy modifications are required for the next 10-20 years, particularly to let breeding work make the desired impact on beneficiaries?

The challenge of the green revolution was total output efficiency, often irregardless of resources. The new challenge is to achieve stable levels of production with economy in resources like land and nutrients, but even more importantly with economy in water use.

Crops capable of reaching above average yields in a given water and nutrient-deficit environment, may be the logical ideotypes of the future.

Agricultural research may be witnessing the beginning of donor fatigue symptoms. Questions are often raised about the impact and benefits of agricultural research, specially in the least developed countries. Statements have been made about the need to move along and assign higher priorities to other research areas and needs.

How can research institutions make a bigger impact on the target communities and environments? Is it possible to enhance crop improvement and particularly seed production and distribution mechanisms? This has been our task during the workshop.

Will the conclusions reached during this meeting have enough merits to categorically state that plant breeding, along with suitable seed production and dissemination schemes, can still make the required difference?

General lessons:

We have learned that there is not a single best mechanism or strategy to achieve success. A large country like India, having a predominant state led system, uses different approaches than a smaller country with a strong private sector like Kenya. Likewise small and grass-roots mechanisms like those used in Latin America necessitate incremental and flexible strategies for expansion.

Open pollinated and self-pollinated crops will also face different requirements.

We have learned that the proper identification of the problem/need (farmers' criteria) is the most critical factor for successful seed production and dissemination mechanisms. The case documented in the Philippines is a good example.

We have learned that excellence in biological research and correct problem/need identification may still not be enough for success. Marketing forces and conditions and the participation of the private sector can make all the difference. The case study from Thailand is an adequate illustration.

We have learned that a team approach, including biological and social scientists, is important for breeding, producing and distributing relevant seed materials.

We have learned that informal dissemination mechanisms of seeds, specially through farmers, are often important and effective.

We have learned that a rigorous testing of improved seeds, under the same conditions where they are supposed to perform, is a critical step.

We have learned that genetic quality by itself is not enough. Post harvest seed management (drying, cleaning, selection, conservation) can be, and often is, as important.

We have also learned that small and marginal farmers do not necessarily assign priority to higher yields. Earliness, tolerance to maladies and stresses, taste, cooking time and other factors are normally the attributes used to adopt or reject new germplasm.

One final word:

Most of the above lessons should help National Research and Development Programs and Projects to design more effective breeding and seed production and distribution systems. Needless to say the same lessons will be equally important for IDRC, and perhaps for other donors as well.

The intense dialogue and interactions achieved during the workshop will need to continue at different levels and different times. Other key elements like plant breeder rights, biotechnology and biodiversity need to be brought into the picture.

SEED PRODUCTION MECHANISMS WORKSHOP
SINGAPORE, 5-9 NOVEMBER 1990
LIST OF PARTICIPANTS

1. Paul M Kimani
Department of Crop Science
University of Nairobi
PO Box 30197
Nairobi
Phone: 592211 Ext 215 (Nairobi)

2. Adriel E Garay
Seed Unit
CIAT
A.A. 6713
Cali, Colombia
Phone: 57-23-675050 Fax: 57-23-647243

3. Neil Thomas
Box 58, RR1
Mallorytown
Ontario, Canada K0E 1R0
Phone/Fax: (613) 659-3807

4. Manee Nikornpun
Department of Horticulture
Faculty of Agriculture
Chiang Mai University
Chiang Mai 50002, Thailand
Phone/Fax: (053) 222101

5. Basudeo Singh
Department of Plant Breeding
G.B. Pant University of Agriculture and Technology
Pantnagar - 263145
Distt. Nainital U.P.
India
Telex: 574261 PANT IN

6. Julieta R Roa
Philippine Root Crop Research and Training Center
Visayas State College of Agriculture
c/o 8 Lourdes Street
Pasay City 1300
Philippines
Phone: 521-20-27 / 58-86-92

7. Faustino Ccama Uchiri
Proyecto de Investigacion de Sistemas
Agropecuarios Andinos
Casilla Postal 388
Puno, Peru
Phone: 350606 Anexo 245 (Office in Lima)
370742 (Home in Lima)

8. Nicolas Mateo
Crop Production Systems Program
AFNS Division
International Development Research Centre
Tanglin PO Box 101
Singapore 9124
Phone: 235-1344
Telex: 21076
Fax: 235-1849

The International Development Research Centre is a public corporation created by the Parliament of Canada in 1970 to support technical and policy research designed to adapt science and technology to the needs of developing countries. The Centre's five program sectors are Environment and Natural Resources, Social Sciences, Health Sciences, Information Sciences and Systems, and Corporate Affairs and Initiatives. The Centre's funds are provided by the Parliament of Canada; IDRC's policies, however, are set by an international Board of Governors. The Centre's headquarters are in Ottawa, Canada. Regional offices are located in Africa, Asia, Latin America, and the Middle East.

Head Office
IDRC, PO Box 8500, Ottawa, Ontario, Canada K1G 3H9

Regional Office for Southeast and East Asia
IDRC, Tanglin PO Box 101, Singapore 9124, Republic of Singapore

Regional Office for South Asia
IDRC, 11 Jor Bagh, New Delhi 110003, India

Regional Office for Eastern and Southern Africa
IDRC, PO Box 62084, Nairobi, Kenya

Regional Office for the Middle East and North Africa
IDRC, PO Box 14 Orman, Giza, Cairo, Egypt

Regional Office for South Africa
IDRC, Ninth Floor Braamfontein Centre, Corner Bertha and Jorissen Streets, Braamfontein, 2001 Johannesburg, South Africa

Regional Office for West and Central Africa
IDRC, BP 11007, CD Annexe, Dakar, Senegal

Regional Office for Latin America and the Caribbean
IDRC, Casilla de Correos 6379, Montevideo, Uruguay

Please direct requests for information about IDRC and its activities to the IDRC office in your region.